64761 35

Frontier Military Series
XXXII

Custer and the 1873 Yellowstone Survey

A DOCUMENTARY HISTORY

Edited by
M. JOHN LUBETKIN

THE ARTHUR H. CLARK COMPANY
An imprint of the University of Oklahoma Press
Norman, Oklahoma
2013

ALSO BY M. JOHN LUBETKIN
*Jay Cooke's Gamble: The Northern Pacific Railroad, the Sioux,
and the Panic of 1873* (Norman, Okla., 2006)

LIBRARY OF CONGRESS CATALOGING-IN-PUBLICATION DATA
Custer and the 1873 Yellowstone survey : a documentary history / edited by M. John Lubetkin.
  pages cm. — (Frontier military series ; volume 32)
Includes bibliographical references and index.
  ISBN 978-0-87062-422-3 (hardcover : alk. paper) 1. Indians of North America—Wars—1866–1895—Sources. 2. Yellowstone River Valley—Surveys—History—19th century—Sources. 3. Railroads—Surveying—History—19th century—Sources. 4. Custer, George A. (George Armstrong), 1839–1876. 5. Dakota Indians—Wars—Sources. 6. Barrows, Samuel J. (Samuel June), 1845–1909. 7. Financial crises—United States—History—19th century. 8. Northern Pacific Railroad Company. I. Lubetkin, M. John.
  E83.866.C959 2013
  973.8'2092—dc23
                         2013016022

Custer and the 1873 Yellowstone Survey: A Documentary History
is Volume 32 in the Frontier Military Series.

The paper in this book meets the guidelines for permanence and durability of the Committee on Production Guidelines for Book Longevity of the Council on Library Resources, Inc. ∞

Copyright © 2013 by the University of Oklahoma Press, Norman,
Publishing Division of the University.
Manufactured in the U.S.A.

All rights reserved. No part of this publication may be reproduced, stored in a retrieval system, or transmitted, in any form or by any means, electronic, mechanical, photocopying, recording, or otherwise—except as permitted under Section 107 or 108 of the United States Copyright Act—without the prior written permission of the University of Oklahoma Press. To request permission to reproduce selections from this book, write to Permissions, University of Oklahoma Press, 2800 Venture Drive, Norman OK 73069, or email rights.oupress@ou.edu.

1 2 3 4 5 6 7 8 9 10

In memory of
John C. Norcutt (1944–2008)
friend and business partner

Contents

List of Illustrations 11
List of Maps 13
Introduction 15

Part I. Preparations

1. THROUGH JUNE 12 41
   Philip H. Sheridan to George W. Cass . . . 45
   W. Milnor Roberts to Cass 47
   Roberts to Thomas L. Rosser and Cass . . . 48
   Montgomery Meigs to His Family 52
   Rosser's Diary, May 22–June 12 58
   Luther Bradley's Diary, May 22–June 13 . . . 60
   Bradley to His Wife 63
   Edward Konopicky to His Parents 65
   Newspaper Articles 69

2. JUNE 13 THROUGH JUNE 17 75
   Samuel J. Barrows to the *New York Tribune* . . . 77
   New York Times 88
   Roberts to Cass 91
   Edward Konopicky to His Parents 94
   Col. David S. Stanley's Memoirs 95

3. JUNE 18 AND 19 97
   Barrows to the *Tribune* 99
   Bradley's Diary and Letter, June 13–19 . . . 105

7

Rosser's Diary and Letters, June 13–19 106
Report of Augustus W. Corliss 108

Part II. To the Yellowstone

4. JUNE 20 THROUGH JULY 7 115
Barrows to the *Tribune* 116
Rosser's Diary, June 20–July 7 133
Rosser to His Wife 136
Bradley's Diary, June 20–July 7 139
Bradley to His Wife 142
Meigs to His Family 144
Sioux City Journal 145
Konopicky to His Parents 147

5. MID-JULY 151
Barrows to the *Tribune* 153
Bradley's Diary, July 8–26 170
Bradley to His Wife 173
Rosser's Diary, July 8–23 175
Phelps's Interview with Custer and Rosser . . . 177
Custer's "Arrest" 179
Konopicky to His Parents 181

6. LATE JULY 185
Barrows to the *Tribune* 187
"Grant's Perilous Ride" 195
Meigs to His Parents 197
Bradley's Diary, July 27–August 3 200
Rosser's Diary, July 24–August 19 201
Charles Loring Brace to the *New York Times* . . . 204

Part III. Fighting on the Yellowstone

7. CUSTER AMBUSHED, AUGUST 4 219
Barrows to the *Tribune* 221
Konopicky to His Parents 233
Bradley's Diary, August 4–10 234

8. **SITTING BULL ATTACKS, AUGUST 11** 237
   Barrows to the *Tribune* 240
   Konopicky to His Parents 256
   Bradley to His Wife 258
   Bradley's Diary, August 11–18 259
   Meigs to His Parents 261
   A. L. Berry's "Remembrance" 262

9. **RETURN TO BISMARCK AND FORT LINCOLN** . . 291
   Konopicky to Dr. Franz Steindachner . . . 293
   Bradley's Diary, August 19 through October 1 . . 294
   New York Tribune 301
   Bismarck Tribune 303
   Rosser's Success 307
   Sioux City Daily Journal 307
   Philadelphia Evening Bulletin 308
   Eckelson's Report to Rosser 309
   Army and Navy Journal 311

Conclusion 313
Selected Bibliography 317
Acknowledgments 321
Index 323

Illustrations

Plates

Pompey's Pillar	209
Watercolor of Little Missouri River, by Charles "Shorty" Graham	209
Little Missouri River	210–211
"Custer and the 1873 Yellowstone Survey," by John K. Ralston	212–213
Scene of August 4, 1873, fighting	212–213
Big Hill from the south	214–215
Approximate location of Custer's wagon train on August 11	214–215
Location of Custer's charge, August 11, 1873	216

Figures

Lt. Col. George A. Custer	265
Samuel J. Barrows	266
Col. David Sloane Stanley	267
Northern Pacific surveyors, 1871 or 1872	267
Luther P. Bradley	268
Montgomery Meigs	268
A. L. Berry	269
W. Milnor Roberts	269
George W. Cass	270
Charles Loring Brace	270
Capt. Augustus W. Corliss	271

William F. Phelps	271
Lt. Col. Frederick Dent Grant	272
Rain-in-the-Face	272
Daniel H. Brush	273
"Heart River," sketch by Charles "Shorty" Graham	273
Maj. Robert E. A. Crofton	274
Joel Asaph Allen	274
Bloody Knife	275
Sitting Bull	276
Gall, or Pizi	277
Norther Pacific Railroad construction	278
Hailstone Creek	279
Entrance to the North Dakota Badlands, near Medora, N.D.	280–281
Eagle's Beak overlooking the Yellowstone near Glendive	282–283
Looking up the Yellowstone near site of August 11, 1873, fighting	284
Steamboat *Josephine*	284
Pompey's Pillar	285
Camp from above	286
Campsite	287
Panning for gold	288
Wagon train	289
Wagon train, Musselshell	290

Maps

Upper Missouri River 22
Yellowstone River 25
Map accompanying Barrows's dispatch in the *New York Daily Tribune*,
   June 28, 1873 76
Tom Rosser's Bismarck to Glendive Creek map, used in 1871 and 1873
   86
1873 Yellowstone Surveying Expedition: Western Dakota and Montana
   184
"Scene of the Battle of Tongue" 218
Custer at the Yellowstone, August 4, 1873 . . . 224
Ambush of Custer, August 4, 1873 227
Custer's position on August 10 and morning of August 11 . 248
Initial Sioux and Cheyenne attacks 249
Stanley's initial arrival and Custer's first moves . . 252
Final Indian attacks; Custer and Stanley's counterattacks . 253

Introduction

C*uster and the 1873 Yellowstone Survey* is the story of the final year's work on the Yellowstone River performed by the Northern Pacific Railroad's engineers. It is also the story of the conflict between the followers of Hunkpapa Sioux leader Sitting Bull and the United States infantry, artillery, and cavalry—these latter units led by George Armstrong Custer—when these forces collided.[1] It is told through the accounts left by the participants, almost all of which were written during the 1873 Yellowstone surveying expedition itself and, with one exception, have not been reprinted since they were written.[2]

During the summers of 1871, 1872, and 1873, five engineering surveys were undertaken by the Northern Pacific into Montana's Yellowstone valley and western North Dakota to establish a final route. While all other *preliminary* surveying had been completed or was under way, the 525-plus miles between present-day Bismarck, North Dakota, and Bozeman, Montana, remained to be surveyed. The reason for postponing this last segment was simple: the land represented the heart of the Sioux Indian hunting grounds and the remnants of the great northern buffalo herds.[3]

In 1873 the Yellowstone Surveying Expedition faced conflict at numerous levels. The survey battled nature's worst summer elements: formidable rain lasting for days, violent thunder- and hailstorms, weeks spent

---

[1] The Hunkpapa were one of the Lakota (Teton) Sioux's seven tribes. Military ranks listed are as of 1873, not the *brevetted* (temporary) rank held during the Civil War and usually by which an officer, as a courtesy, was addressed.

[2] Rosser's 1873 diary and letters were published by this editor in *North Dakota History* 70 (Fall 2003).

[3] Cheyenne, Crow, Nez Perce, Métis, and others also hunted in this ecologically imploding area.

in broiling heat under cloudless skies with exceptional humidity, and all made nastier by the constant buzzing of insects. Adding to everyone's burden was the psychologically debilitating squabbling between infantry and cavalry officers, at one point turning sufficiently nasty to cause Custer's "arrest." The cavalry officers themselves were sharply divided into pro- and anti-Custer camps. Making matters worse, the expedition was led by the crusty but able Col. David S. Stanley, who at forty-five not only didn't want to be in the field for a second year in a row but sadly lived up to his reputation for binge drinking.

Not to be forgotten is the psychological undercurrent that, after crossing the Missouri, virtually every person—military and civilian alike—lived in dread of being captured, tortured, and mutilated. Maj. Robert Crofton, a respected wartime combat officer, was no exception. That fall he wrote, "The fear of red skins was before your eyes all the time." This constant dread was magnified by rumors and exaggerated newspaper reports of thousands of Indians waiting to attack.[4]

Yet despite Indian fighting, internal conflict, and frequent boredom, the surveying expedition was not devoid of humorous, often slapstick incidents: a hundred soldiers simultaneously falling on their faces into a foot of mud, an officer climbing onto his chair as his tent filled with water, a significant army snafu that no novelist could invent, and a mule that apparently fell in love with an Indian pony.

By summer's end Custer's reputation as the army's most dashing and able cavalry officer was reaffirmed. His victories were so decisive that the Sioux would not attack army troops again until tactically forced to do so in 1876. Additionally, 1873's fighting—miniscule by Civil War standards—helped trigger the second worst depression or recession in American history.[5]

Jay Cooke and the Northern Pacific

The Northern Pacific, as Congress named the railroad in 1864, was a concept that its critics claimed ran from nowhere to nowhere. In the years following the Mexican War, numerous transcontinental railroad

---

[4]Letter from Maj. Robert E. A. Crofton, October 28, 1873, in Lubetkin, *Jay Cooke's Gamble: The Northern Pacific Railroad, the Sioux, and the Panic of 1873* (Norman: University of Oklahoma Press, 2006), 321n24.
[5]Paul Krugman, "The Third Depression," *New York Times*, June 27, 2010 (op-ed page).

routes from the Mississippi to the Pacific were put forward by regional interests. Congressional support for any route was a financial necessity, but like so many issues in the 1850s, Northern and Southern interests collided, resulting only in the *study* of five possible options.

The Civil War changed everything. The North needed to safeguard California and Nevada's gold and silver. Southern opposition to central and northern routes was gone, and railroad and related technology (e.g., dynamite for construction) was rapidly advancing. Congress passed the 1862 Pacific Railroad Act to build the Omaha to Sacramento Route, and in May 1869 the east and west coasts were linked by rail.[6]

Yet even as word of this was telegraphed around the country, planning for a second road was under way. But was there really an economic *need*, and what, if anything, was the federal government's obligation? Certainly there was ample precedent for government assistance. Since 1827, to stimulate interstate commerce, the government frequently had provided cash, land grants, and bond guarantees to build canals and railroads. Yet the truthful business answer to the question of need for the Northern Pacific was no. Testifying before Congress that year, Gen. William T. Sherman stated that if the just-completed transcontinental were to be double-tracked, there would be no need for a second east-west road for fifty years. Wrong as Sherman was about the fifty years, in the decade after the Civil War there was no demonstrable economic need for a more northern transcontinental railroad. Additionally, the overwhelming number of Americans (already 800,000 and growing by 45,000 a year) settled in California, Nevada, Utah, and Nebraska lived within one hundred miles of the transcontinental track.

The 1864 Pacific Railroad Act creating the Northern Pacific was not a visionary idea whose time had come as much as a testimony to effective lobbying. While most of the act concerned the first transcontinental railroad, the Northern Pacific's section cost the public nothing—if 50 million acres of land grants isn't counted—and mollified Northern tier representatives. During its first five years, the men owning this paper railroad raised $200,000, performed modest surveying in Minnesota, but laid not an inch of track.

In 1869 the Northern Pacific's charter was nearing the end of its congressional mandate to undertake construction when a decision by

---

[6]The exception was Council Bluffs, Iowa, where a ferry took passengers over the Missouri to Omaha.

just-inaugurated President Ulysses S. Grant changed everything. After taking office Grant declined to nominate a logical choice, Jay Cooke ("the Financier of the Civil War"), to be his secretary of the treasury. Cooke, in many respects the second most powerful man in the country, owned the nation's largest bank. He was revered by the public for his scrupulous honesty in raising more than a quarter of the funds needed to finance the North's victory. Cooke, shocked by Grant's decision, never recognized that his own idiosyncrasies were to blame. As he literally believed himself to be "God's chosen instrument," he thought that if Grant refused to select him, there had to be a *Reason*. That reason, Cooke revealed in his unpublished memoirs, was to build the Northern Pacific, a route Cooke suddenly "understood" he had been preparing for his entire life.[7]

On January 1, 1870, the Northern Pacific Railroad and its banker, Jay Cooke & Co., formally signed an agreement that would create Yellowstone Park, construct the nation's second transcontinental railroad, motivate the Canadian and British governments to build the Canadian Pacific Railroad, attract a flood of Scandinavian settlers to Minnesota and Dakota, and reignite war with the Sioux. The railroad was no small achievement, though its implementation would be interrupted by the devastating Panic of 1873, caused in large part by the failure of Cooke & Co. Yet despite the economic depression that followed, in another decade two transcontinental railroads (the Northern Pacific and the Southern Pacific) would be completed, bringing unprecedented economic growth to the nation.[8]

In sharp contrast to California, Nevada, Utah, and Nebraska's demographics, the Northern Pacific's planned route passed within one hundred miles of perhaps 110,000 settlers in Oregon, Washington, Idaho, Montana, and Dakota and with combined growth of, at best, 7,500 annually. If the numbers were weak, between Walla Walla, Washington, and Brainerd, Minnesota, virtually no *railroad* surveying had been undertaken. Numerous passes along the Continental Divide were known only to Indians and illiterate frontiersman. Army surveys had

---

[7]Lubetkin, *Jay Cooke's Gamble*, chapter 1, "God's Chosen Instrument." For a far different picture of Cooke, see Richard White, *Railroaded: The Transcontinentals and the Making of Modern America* (New York: Norton, 2011).

[8]For brief biographies of Cooke and Smith and NPRR history, see Robert L. Frey, ed., *Railroads in the Nineteenth Century* (New York: Facts on File, 1988).

been undertaken across the northwestern portion of the United States in the 1850s to accurately show rivers and mountain ranges, not railroad routes. Indeed, the primitive understanding of the area's geography is best illustrated by Yellowstone Park itself, which remained "undiscovered" until 1869.[9]

The needs of Montana's population of approximately 20,000 people were served by oxen-drawn freighters from Salt Lake City and Missouri River steamers to Fort Benton. Wagons and steamers carried food, clothing, arms and ammunition, liquor, and printing presses—every community with 1,000 people seemed to have a weekly newspaper—as well as smelters, milling, and mining equipment. As for Dakota, most of its 7,500 whites lived in the southeastern part of the territory, hundreds of miles from the Northern Pacific's planned route.

Nevertheless, Cooke decided to push ahead. It was his instinct that created Yellowstone Park, and he was equally enthusiastic about western settlement, particularly the fertile Red River valley. What he knew about Minnesota's mineral deposits is unclear, but it was Cooke's decision to utilize the port of Duluth. And, to assist the Northern Pacific, Cooke was never hesitant to bribe congressmen, break Indian treaties, or flout public waterways law.[10]

After the war, Cooke seldom made significantly profitable investments. His five homes—including the 70,000-square-foot *Ogontz* and one on a Lake Erie island he owned—satisfied his needs, but he had become bored; making money for its own sake had never appealed to him. Not yet fifty, he needed to sink his teeth into something and to be challenged as he had been during the war.

Cooke's greatest talent was selling bonds, and his skills lay rooted in his belief in America and its miraculous growth. As a teenager he'd been a clerk in St. Louis and watched some of the first steamers churn up the Missouri for eastern Montana. Later Cooke helped finance midcentury American infrastructure, especially canals, railroads, and telegraph companies. Cooke had always bet on "progress," and he had been inevitably proven correct.

---

[9]The exception was the study by Isaac I. Stevens. Little of this effort was used by the Northern Pacific, as its line was built south of Stevens's route. However, parts were incorporated into the Mullen *wagon* road.

[10]Evading the Army Corps of Engineers and court orders, in 1871 Cooke had a cut built through a narrow, sandy peninsula point, thus opening Duluth for shipping.

Settlement and the Sioux

Late in 1862, gold was discovered in Idaho. Then, in rapid succession came additional Idaho and Montana gold and silver strikes, many of them gigantic. With large numbers of Confederate sympathizers and deserters in Montana, the necessity for safeguarding the gold became an added concern for the Lincoln administration.

Coincidentally, just months before the Idaho strike, half-starved, poorly armed Sioux living on a squalid Minnesota reservation staged an "uprising" that, after some initial successes, petered out. With this as an excuse, in the summers of 1863–65, the army, bolstered by Minnesota, Iowa, and Colorado frontier regiments, attacked the central and western Sioux and Cheyenne mostly living or hunting east of the Missouri. In the ensuing campaigns, the army inflicted a series of stinging defeats at Big Mound, Dead Buffalo Lake, and Whitestone Hill (all in 1863) and Killdeer Mountain (1864). These grievous losses, combined with the sudden decline of buffalo in Dakota (likely from disease), left the remaining Sioux, mostly Lakota, badly shaken, their morale and confidence all but broken. Despite being technical winners of the Bozeman Trail fighting, by 1871 a majority of Lakotas, led by Red Cloud, were forced to live on reservations (then called *agencies*), dependent on annual government handouts for their very survival.[11]

Nevertheless, a minority of the overall Sioux population—mostly Lakota but with some Dakota and Santee—and a handful from other Plains tribes followed the "militant" leadership of Sitting Bull, a Hunkpapa of the Lakota (or Teton) Sioux tribe. These Indians, whose winter numbers never exceeded a thousand lodges, chose to live in their traditional hunter-gatherer way, wintering in northeastern Montana. By 1870, as Sitting Bull's adherents struggled to survive, they became engaged in a form of ecological Russian roulette by killing buffalo rapidly and trading the hides for repeating rifles and ammunition.

Mismanagement on a Grand Scale

The January 1870 Cooke–Northern Pacific goal was to reach the Red River (present-day Fargo, North Dakota, and Moorhead, Minnesota)

---

[11]For an excellent account, see Micheal Clodfelter, *The Dakota War: The United States Army versus the Sioux, 1862–1865* (Jefferson, N.C.: McFarland, 1998). For a list of Indian agencies, reservations, and troops in 1870–73, see *Jay Cooke's Gamble*, 50.

by July 1, 1871. However, not until August 1870 did actual construction begin.[12]

Unfortunately, building a railroad two thousand miles in length required certain skills and experience that neither Northern Pacific president J. Gregory Smith nor his small management team possessed. Worse, however, was that Smith's "Vermont Clique"—now Jay Cooke's partners—was, in this writer's opinion, corrupt and conflicted. And even if they had been scrupulously honest, Smith's team would still have been too small, overconfident, and poorly led.

Smith, competent and hardworking, had single-handedly developed the Vermont Central Railroad into the country's ninth largest rail system. However, his business résumé included numerous ethically questionable dealings, to say nothing of his inability to meet competition or economic downturns. Smith's business-mature railroad also faced competition from the midsized and Vermont-based Rutland Railroad, which was also operating in a stagnant market. Smith's competitive answer was to "lease" the Rutland (1870), and from the first he was unable to meet his payments.

As the Northern Pacific's president, Smith placed its headquarters in New York City, though he was seldom there. Sadly, he was to be found even less frequently in Minnesota, the hub of the Northern Pacific's construction efforts. Despite some preconstruction surveying, Smith apparently was caught flatfooted by the Northern Pacific's problems in its first significant segment: 125-plus miles of track through Minnesota wetlands. Glaciers had flattened this area, leaving it with meandering streams, swamps thirty feet deep, floating islands, thousands of lakes, mosquitoes beyond belief, and ferocious winters, the coldest in the "lower 48."[13]

With inadequate equipment (e.g., only one pile driver), a route motivated by land speculation, and the decision to build *over* ice during 1870–71's winter, it wasn't until June 1872 that the Red River was bridged at Fargo–Moorhead, eleven months behind schedule. From the first, Smith purchased excessive equipment (especially locomotives), awarded questionable contracts to unknown companies, built unnecessary track,

---

[12] The Pacific Railroad Act required the Northern Pacific to build east to west *and* west to east. This latter effort, utilizing perhaps one-eighth of the company's funds, is not discussed here.

[13] For details of the railroad's Minnesota problems, see John L. Harnsberger, *Jay Cooke and Minnesota: The Formative Years of the Northern Pacific Railroad, 1868–1873* (New York: Arno Press, 1981).

The Upper Missouri River

The economic importance of the Northern Pacific's reaching the Missouri at Bismarck meant that some 750 river miles and weeks of steaming against the current could be eliminated from Sioux City, Iowa. For fifteen years, river traffic continued up the Missouri for another 1,055 river miles (as measured in the 1870s) from Bismarck to Fort Benton. *Copyright © 2013 by The University of Oklahoma Press.*

excessive facilities (*two* Brainerd stations in eighteen months), and spent Cooke's money extravagantly. By mid-1872 the railroad and Cooke were out of cash. Worse, strategic construction, especially from St. Cloud to Brainerd, remained uncompleted, most likely to protect insider real estate holdings in Duluth. Even as late as May 1872, a month before Smith "resigned," he and his allies believed Cooke's financial warnings were simply bluffs.

The 1871 Surveys

Despite the growing construction and financial problems, preliminary engineering under the overall direction of chief engineer Milnor Roberts, an able and honest executive, made excellent progress. Thus in September 1871, Northern Pacific's engineers cautiously pushed into the Yellowstone valley to map the area and test Indian reaction.

Leaving Bozeman and Fort Ellis, Montana, a score of engineers and two cavalry companies (some one hundred officers and men) crossed the Bozeman Pass and reached the Yellowstone valley at present-day Livingston. For one hundred miles east, the valley was nominally controlled by friendly Crow Indians. Moving slowly, the survey proceeded to present-day Billings. While the work proved physically difficult, it was otherwise uneventful until, on returning, the column was engulfed by a November blizzard.[14]

Meanwhile, a second survey left Fort Rice (25 miles south of today's Bismarck) on September 9. Marching west, some five hundred men followed the 180-mile Heart River to its source near current Belfield, North Dakota. This column survived a raging prairie fire and marched through biting snow, but other than the alcoholic commanding officer's inability to read a compass, it proved successful. A gap in the Badlands along the Little Missouri (present-day Medora, North Dakota) provided a relatively straight railroad route between Bismarck and the Yellowstone. This route would be perfected over the next two years and remains in use today, roughly paralleled by Interstate 94.[15]

Meanwhile Indian hostility, predicted by the railroad's business enemies, never materialized. In fact, not until both surveys were well under way did Sitting Bull learn of them. Uncertain as to their

---

[14]*Jay Cooke's Gamble,* chapter 7, "Stop Firing at Me."
[15]Ibid., chapter 8, "The Army of the Glendive."

purpose—combined with placating gifts from knowledgeable Indian agents—Sitting Bull chose to ignore them. However, during the winter, he discovered not only their purpose but also that two additional surveys were planned. Fully aware of the consequences of a railroad in the Yellowstone valley, in April Sitting Bull sent an emissary, Spotted Eagle, to meet with Colonel Stanley at Fort Sully, present-day Pierre, South Dakota. Spotted Eagle's efforts to quash the forthcoming surveys were predictably futile. Stanley noted that the Indians said that "driving away the buffalo was [their] death" and that they "would tear up the railroad and kill the builders." As gently as possible Stanley told Spotted Eagle that the railroad was going to be built, period. Spotted Eagle's final remarks were despairing: the Indians would fight, he said, although they knew they would be killed in the process.[16]

Critically important to Northern Pacific efforts was the backing of the Grant administration and Lt. Gen. Philip H. Sheridan, commander of the Division of the Missouri. This division, by far the army's largest, had approximately 16,500 officers and enlisted men, half the army's peacetime strength of 32,500. Sheridan viewed the Northern Pacific as supporting his own Indian pacification goals and, with Grant's tacit approval, ran the division all but independently of his superior, Gen. William T. Sherman.[17]

Following Spotted Eagle's failed efforts, Sitting Bull sent emissaries to numerous Plains tribes, calling for an August rendezvous on the Powder River near the Montana–Wyoming territorial line. At least 2,000 Sioux, Cheyenne, Arapaho, and other warriors (and often their families) heeded the call, and as they deliberated, the two surveying expeditions left Forts Rice and Ellis.[18]

The 1872 Surveys

The western column consisted of four companies each of infantry and cavalry, just over 375 officers and men and, all told, approximately 525 participants. Commanding was Maj. Eugene M. Baker, a wartime

---

[16]Robert M. Utley, *The Lance and the Shield: The Life and Times of Sitting Bull* (New York: Henry Holt, 1993), 107.
[17]Annual *Army Register* and *Report of the Secretary of War* to Congress.
[18]Utley, *The Lance and the Shield,* 107, and Kingsley M. Bray, *Crazy Horse: A Lakota Life* (Norman: University of Oklahoma Press, 2006), 161.

The Yellowstone River

The Northern Pacific Railroad spawned numerous communities along the Yellowstone, including Glendive, Miles City, and Billings. Perhaps surprisingly, no communities grew where the Bighorn or Powder River flowed into the Yellowstone. In 1875 the *Josephine* traveled as far upstream as Billings, some 460 river miles above Fort Union, (North) Dakota, where the Yellowstone and Missouri met. *Copyright © 2013 by The University of Oklahoma Press.*

combat officer well liked by Sheridan and a popular Montana figure. Baker, often in his cups, was best—or infamously—known for leading the 1870 Marias Creek Massacre in which some 170 Blackfeet died while he suffered zero casualties.

After Baker crossed Bozeman Pass, he found it impossible to ford the still-high Yellowstone. Rather than wait for the river to drop, Baker continued along its north bank. East of the Crazy Mountains, the column took to the difficult highlands between the Yellowstone and Musselshell, reaching Billings's Rimrock area on August 12. Baker, over a week and a hundred miles behind schedule, still had to cross the Yellowstone. When his column found a lush, idyllic campground on the river, they collapsed. No attempt was made to move on August 13, and despite numerous Indian signs, Baker felt in no danger. He posted a minimal, less than vigilant guard, and drank himself insensible that evening.[19]

While most of Baker's column slept, perhaps 1,000 Sioux, Cheyenne, and Arapaho were approaching his camp. A conference was held, and while no attack appears to have been planned, impatient young warriors started to independently swim the Yellowstone. Gradually they surrounded Baker's camp in preparation for—at a minimum—stealing his horses. One particularly restless warrior, with the improbable name of Plenty Lice, sneaked *through* the camp's outskirts and about 3 A.M. tried to steal a rifle. In the process Plenty Lice was shot and the fight began.[20]

Fortunately for the soldiers, a number of veteran officers were up and still playing poker. Although Baker was too drunk to be awakened, the officers took the lead, woke and rallied the troops, and soon pushed the Indians back to a nearby ridge. By the time the fighting broke off, the Indians had sustained heavier losses, but both sides' combined dead and seriously wounded was under a dozen. Later the officers jocularly named the fighting the "Battle of Poker Flats," a takeoff on Bret Harte's popular short story "The Outcast of Poker Flats."

However, the action left Baker unnerved. Rather than continue, he undermined the surveyors' attempts to continue working by refusing to move his campsite and thus forcing the surveyors and their fifty-man guard to work further and further from camp each day. Next Baker

---

[19] Baker's inebriated state is described by Lt. James A. Bradley in his *March of the Montana Column: A Prelude to the Custer Disaster*, ed. Edgar I. Stewart (Norman: University of Oklahoma Press, 1961).

[20] *Jay Cooke's Gamble*, 140.

maneuvered the railroad's frightened lead engineer into making a written request to return to Fort Ellis. Opposite Pompey's Pillar, fifteen miles downstream, the column retreated north to the Musselshell.

With 150 miles between the two surveys unmapped, both became failures, although it could have been far worse. One can only speculate on the what-ifs had the overconfident Baker crossed the Yellowstone. With only two Indian scouts, one the legendary Mitch Bouyer (or Boyer), Baker could have easily walked into an ambush by 1,000 warriors, by far a bloodier outcome than "Poker Flats."[21]

Disconcertingly, the Indians appear to have been fully aware of the Northern Pacific's engineering goal and correctly credited themselves with a significant victory. Having displayed their bravery and used up too much ammunition, most of them promptly deserted Sitting Bull and went hunting for the remainder of the summer. This left the Hunkpapa leader with perhaps two hundred warriors with which to fight Colonel Stanley.

Stanley at Bay

Stanley's force consisted of 535 infantry officers and men, supported by three Gatling guns and one Napoleon twelve-pounder artillery piece. Despite his request, he was given no cavalry, leaving him without any offensive punch against mounted Indians. Leading the surveyors for the second year in a row was an ex-Confederate cavalry major general, Thomas L. Rosser, an engineer of indifferent technical skills but a marvelous leader, motivator, and military tactician.

Supported by good weather, Stanley's column made rapid progress and reached Montana's O'Fallon Creek valley in under three weeks. Although they'd been spotted by Hunkpapa Sioux scouts under Gall, not until August 16 was any aggressive move made. That dawn, Gall and twenty or so warriors galloped out from some trees and headed directly for sleeping surveyors, sending a volley of arrows and bullets before dashing away. The morning's only fatality was an army mule shot by the surveyors which, an hour later, was eaten by Gall's starving warriors.

Soon, however, Gall's warriors boasted—"trash talk" in today's idiom— to Stanley's Indian scouts about their victory at Poker Flats two days

---
[21]Ibid., chapter 10 and maps, "The Battle of Poker Flats."

earlier. Stanley thus became aware of the fighting, Baker's retreat, and Sitting Bull's plans to join Gall, a threat not to be taken lightly. Within days, Sitting Bull arrived, but his efforts to "punish" Stanley proved unsuccessful. A poorly executed ambush, a prairie fire set too early in the morning (it didn't ignite because of dew), and a halfhearted charge against an ineffective Gatling gun proved no match for Stanley's well-disciplined infantry.[22]

Sitting Bull and Gall quickly disappeared. For the following four weeks, the surveyors unsuccessfully searched for a railroad shortcut between the Yellowstone and Bismarck, hoping to eliminate fifty miles of track construction with a corresponding financial savings. On October 1, the column, now at the Little Missouri, turned back to Fort Rice. By now the men were low on water, unable to bathe, eating salted pork, walking in shoes ("brogans") with *cardboard* soles, and sleeping in leaky tents. Their condition was so bad that one officer characterized his infantrymen as "Falstaff's ragamuffins."[23]

Could anything worse happen? The answer, of course, was yes. Proving the old adage that no good deed goes unpunished, Stanley split his forces on October 1 so that three infantry companies with rotted tents might reach Fort Rice before real cold set in. Unbeknownst to Stanley, Gall and some fifty Hunkpapas were shadowing them. On October 3, a one-armed officer with this smaller column, Lt. Eban Crosby, went hunting alone. Soon he was surrounded by Gall's warriors, who killed, mutilated, and scalped him. Sitting Bull biographer Stanley Vestal wrote: "It is said that Gall collected his remaining arm and took it home as a trophy."[24]

Curiously, Crosby's senior officer, Maj. Robert Crofton, did not notify Stanley, later claiming he was afraid to risk a messenger's life. Thus when Gall struck again on October 4, Stanley's column was caught unawares. In separate ambushes Gall killed Stephan Harris, Stanley's

---

[22]Ibid., chapter 11, "Falstaff's Ragamuffins"; David Eckroth, Howard Boggess, and Mike Penfold, *Sitting Bull's Fight with Colonel Stanley, August 22, 1872: "The Battle of O'Fallon Creek,"* American Battlefield Protection Program, March 2010, GA-2255-08-005. Eckroth was also one of the discoverers of the "Poker Flats" site.

[23]One officer, using the name "Nisbey Swipes," wrote regularly to the *Sioux City Daily Journal.* The Falstaff comment comes from the October 19 and 26, 1872, issues.

[24]Stanley Vestal, *Sitting Bull: Champion of the Sioux* (Boston: Houghton Mifflin, 1932; reprint, Norman: University of Oklahoma Press, 1957), 131.

black servant, and almost trapped Rosser. In chasing Rosser, the Indians mortally wounded an officer who was hunting alone on foot, Lt. Louis Dent Adair, the first cousin of First Lady Julia Dent Grant.[25]

For Phil Sheridan and the Grant family, Adair's death was an added slap in the face after Baker's humiliating retreat. Indian pressure had caused the expensive surveys to fail, and the casualty list reached into the nation's First Family. Adair and Crosby's deaths caused a huge stir in the national press, in part because they came in the waning days of a presidential campaign.

Soon the president's son, Frederick Dent Grant, age twenty-two, an army officer assigned to Sheridan's staff, announced that he would go on the following year's survey, in effect avenging the cousin that he possibly had never met.

Of far greater importance, Cooke's bank, the nation's largest and seemingly safest, was overextended, and Cooke was performing a high-wire financial balancing act. Worse for the railroad, Northern Pacific bond sales evaporated. A second source of income also collapsed when no federal bonds were issued in 1872, disastrously impacting on Cooke's bank. Finally, of Minnesota's three million acres of federal land to be transferred to the Northern Pacific, Smith refused to file for the requisite permits, including tens of thousands of exceptionally fertile acres. Not until late fall 1872 did the railroad receive federal permission.[26] By then Smith was gone, replaced by Gen. George Washington Cass, an experienced railroad executive.

Sheridan was soon planning for 1873, and somehow Cooke, his bank, and personal friends raised additional funds to complete the forty miles of track to Bismarck—an economic necessity for the railroad. Meanwhile Sheridan, now aware of the railroad's problems, began throwing army contracts to the Northern Pacific. Equally important, Cooke wanted to finish, once and for all, the line of survey.

---

[25]M. John Lubetkin, "An Ambush That Changed History: Thomas L. Rosser, David S. Stanley, Gall, and the 1872 Eastern Yellowstone Surveying Expedition," *Journal of the West* 42 (Winter 2004): 74–83.

[26]See *Jay Cooke's Gamble*, chapter 12, and Clarence G. Glasrud, ed., *Roy Johnson's Red River Valley* (Moorhead, Minn.: Red River Valley Historical Society, 1982). Why didn't Smith file? In 1871, government certification would have exposed the shoddy construction. Also, Smith owned (whole or in part) significant Minnesota acreage, and selling Northern Pacific land would have put him in competition with himself.

George Armstrong Custer

It is here that this book begins, a story told almost entirely by the surveying expedition's civilian participants and, with few exceptions, written in 1873 or 1874. The importance of contemporary accounts cannot be overstated because, following Custer's death, virtually everything concerning him was colored by the Little Big Horn disaster. For those particularly interested in Custer and how he was viewed by his civilian contemporaries (Barrows, Konopicky, his old friend Rosser, Phelps, Lounsberry, Meigs, and others), this book will offer a previously unseen and—to let the cat out of the bag—frequently flattering portrait.[27]

In 1873 the mere word "Yellowstone" invoked a sense of scientific mystery to the public: Were further miracles of nature waiting to be found? Would new "petrifications" (dinosaur fossils) be literally unearthed, and if so, what impact might they have on Darwin's controversial theories? And, of course, did *yellow-stone* mean that large gold deposits remained undiscovered?[28]

Further whetting public anticipation was Sheridan's decision to give Stanley an offensive punch in the form of the Seventh Cavalry commanded by the often combat-hapless Col. Samuel D. Sturgis. Sheridan, however, placed the Seventh's headquarters in Fort Snelling (St. Paul), Minnesota, and ordered Sturgis and his staff there while sending ten of the regiment's twelve companies, or 450 men, under the command of its lieutenant colonel, George Armstrong Custer, to Fort Rice.[29]

Custer needs no introduction, and this writer suspects that 99 percent of this book's readers have formed their opinion of him. His life and death have been portrayed in thousands of books, articles, monographs, movies, TV shows, Wild West re-creations, paintings, sculptures, and so on. With the possible exception of Gettysburg or D-Day, no battle is better known to Americans than the 1876 annihilation of Custer's personal command.[30]

---

[27]In 1906 a surveyors' reunion was held. Afterward, numerous letters were received with invaluable information found nowhere else.

[28]"20,000 Indians," *New York Herald*, September 12, 1872.

[29]Samuel D. Sturgis (1822–89), West Point '46. Sturgis, while brave enough, was captured in the Mexican War, again in the Civil War, suffered one of the North's worst defeats at the hands of Nathan Bedford Forrest at Brice's Cross Roads, and to be defeated by Chief Joseph in 1877.

[30]Biographies of Custer are too numerous to cite. Two that this writer found particularly useful were Jeffry D. Wert, *Custer: The Controversial Life of George Armstrong Custer* (New York: Simon and Schuster, 1996) and Robert M. Utley, *Cavalier in Buckskin: George Armstrong Custer and the Western Military Frontier* (Norman: University of Oklahoma Press, 1988).

INTRODUCTION 31

Today it remains fashionable to assume that Custer was stupid; after all, the argument goes, he *was* last in his West Point class and *did* get himself killed. However, within a year of his West Point graduation he had served on the staff of tough-minded Gen. Phil Kearney, was one of the first officers to ride in an observation balloon, and was cited for bravery by Winfield Scott Hancock. After one particularly gallant action, two generals suggested to George B. McClellan that he meet Custer. McClellan did just that, immediately liked him, praised him for his modesty, and on May 28, 1862, formally added Custer, just twenty-two years old, to his staff.

McClellan's staff was not for lightweights. Whatever McClellan's combat limitations, he had one of the best organizational minds in the country, and his team included members of some of the country's most renowned families, including an Astor and a Biddle. Yet here Custer shone, although he was penniless, without influence, and burdened by a West Point reputation to live down. Custer's unique ability to "see" and then rapidly and accurately *describe* battlefield conditions was invaluable for any commanding general. While this ability might appear natural, only Grant and a handful of others possessed it. After Antietam, when McClellan was sacked, senior officers maneuvered to place Custer on their own staffs.[31]

On taking field command of four regiments (some 1,300 cavalrymen) as a brevetted (temporary) brigadier general, Custer was like a public relations magnet, his combat achievements becoming known throughout the country. As one writer recently noted, he became one of the "first self-made American celebrities."[32]

On June 30, 1863, at Hanover, Pennsylvania, east of Gettysburg and days after his appointment, Custer blocked J. E. B. Stuart's access to Lee, forcing Stuart to swing north to Harrisburg and Carlisle. July 3 saw Custer prevent Stuart from reaching the Union army's vulnerable supply train. In the following spring, Custer's "Wolverines" mortally wounded Stuart. That fall Custer gained a major general's second star. At Appomattox, Custer captured trainloads of supplies meant for Lee and blocked the retreating Confederate army.

---

[31]Ibid., chapter 3, "A Beginning?"

[32]See Bruce Barcott's *New York Times* review of Nathaniel Philbrick's *The Last Stand* (June 13, 2010). However, Walter Isaacson argues rather persuasively in *Benjamin Franklin: An American Life* (New York: Simon and Schuster, 2003) that it was this Founding Father who was the country's first great celebrity.

Here, however, Custer made the second worst mistake in his life. Meeting with Gen. James Longstreet (Grant's West Point classmate and friend) under a flag of truce, Custer asked Longstreet to surrender to Sheridan, not to Grant, a significant violation of military protocol. Custer's impetuous, ill-conceived request resulted in a "black mark" from Grant, one that Custer never fully overcame, despite Sheridan's continuing support.[33]

After the war, Custer acquired a reputation for Indian fighting, harsh leadership, and causing trouble for immediate superiors. After almost a year's suspension for deserting his troops—for conjugal reasons—he was yanked back by Phil Sheridan, who wanted him for a special mission on the Kansas frontier. November 27, 1868, saw Custer launch an early morning attack on a medium-sized Cheyenne village. Whether the fighting was a "massacre" (more women and children appear to have died than warriors), whether the village deserved to be attacked, and whether Custer "abandoned" Maj. Joel H. Elliott and his small command (seventeen men dying) all remain contentious issues. Certainly the fighting in the Indian village unleashed a "firestorm of protest from critics of the army," with Custer and Sheridan called "savages." While Sheridan brushed aside the newspaper denunciations, Custer appears to have been stung by the criticism, and this might explain his slow pursuit of Sitting Bull prior to the fighting on August 11, 1873.[34]

Custer spent 1869–70 on the Kansas frontier and then languished for two years on Reconstruction duty in Kentucky; his responsibilities were whittled down to two companies, a fraction the size of his 1865 command. To offset his modest salary and boredom, Custer turned to magazine writing, proving himself a graceful wordsmith while keeping his name before the public.

David Sloan Stanley

Custer's commander in 1873 was forty-five-year-old David Sloane Stanley, colonel of the Twenty-second Infantry Regiment and a Northern Plains veteran. While no officer was as close to Sheridan as Custer, in 1848 Stanley and Sheridan traveled together from Ohio to West Point

---

[33]Wert, *Custer*, 224–26.
[34]Ibid., 278–79. Also see Nathaniel Philbrick, *The Last Stand: Custer, Sitting Bull, and the Battle of the Little Bighorn* (New York: Viking, 2010), 12–13, 133–37.

during their plebe (freshmen) year, forming a special bond that salvaged Stanley's career on more than one occasion.

Stanley was intelligent (ninth in his West Point class), disciplined, sensitive, and logical. He was also withdrawn, moody, and introspective, in many ways an outsider, with few close army friends. His mother died when he was young, and after his father's remarriage he was, at twelve, apprenticed to a doctor. Until his death he remained conflicted between the army and his desire to practice medicine. Tellingly, he chose to marry the daughter of the army's leading physician, Joseph J. B. Wright. Despite his impoverished childhood, Stanley was not reluctant to own slaves and likely because of this was informally offered a position in the Confederate army.[35]

During the war Stanley proved personally brave, excelling in tactical defense and the use of field artillery. He was respected by Grant and became George H. Thomas's protégé. However, during the final phase of the 1864 Atlanta campaign, as Corps commander (replacing the popular James B. McPherson, who had just been killed), Stanley failed to complete the encirclement of Hood's army, incurring Sherman's enduring wrath. And, to some degree, Stanley was certainly at fault for not being sufficiently aggressive.[36]

By 1867 Stanley found himself a colonel at isolated Fort Sully, a Missouri River post in south-central Dakota. Here he began drinking. Alcoholic symptoms vary, but Stanley proved to be a binge drinker. To what degree this impacted on routine army life is unknown, but various Northern Pacific surveyors indicate he lost control of himself on three occasions in 1872. What characterized his binge drinking was that it invariably followed periods of intense stress. In short, he was always aware when it was *safe* to let himself go. Despite this, Stanley's overall leadership was solid; his infantry performed well, and he fully supported the Northern Pacific's efforts, something that cannot be said for the three other 1871 and '72 surveys.

---

[35]This and subsequent paragraphs are primarily from *Personal Memoirs of Major-General D. S. Stanley* (Cambridge: Harvard University Press, 1917; reprint, Gaithersburg, Md.: Olde Soldier Books, 1987). In chapters about his Ohio boyhood Stanley proves a graceful writer, but later sections become dense reading, often marred by gratuitously nasty comments, especially about Sheridan.

[36]For the Atlanta campaign and the Sherman-Stanley feud, see Albert Castel, *Decision in the West: The Atlanta Campaign of 1864* (Lawrence: University Press of Kansas, 1992), especially appendix C.

Stanley is often portrayed as constantly in his cups, responsible for delay after delay and endangering his command in 1873. This is simply inaccurate. Stanley was *possibly* drunk one night near the end of the first week and *definitely* went on a shocking three-day July binge while camped on the Yellowstone. However, this paralleled a week's delay caused by the delivery of all the army's supplies to an unreachable location—the snafu alluded to earlier. At most, a single day's delay was lost to Stanley's drinking. More importantly, when Custer twice faced off against Indians in August, on both occasions Stanley arrived with reinforcements on a timely basis. Yet, as people who have worked for alcoholics are aware, there is always a sense of mistrust and worry. What would Stanley do when, not if, the Indians attacked? Although Stanley made no military errors and his route of march was selected by surveying requirements, his leadership had to have been a psychological burden on all participants.[37]

Chroniclers of the 1873 Survey

The reader will quickly discover that the heart of this book is the account of the 1873 survey written in a series of letters by Samuel June Barrows and sent to the *New York Tribune*.

A hundred years after his death, Barrows (1845–1909) is almost forgotten, but his life, beginning in abject poverty in New York City, can only be described as astonishing. He was just three when his father died, and his mother eked out a marginal existence manufacturing bootblack. This she made at home, the key ingredient being molasses. One thinks of a Dickens-like setting as her five children, inhaling noxious fumes, struggled to mix, heat, and stir the thickening concoction until it could be poured into small molds.[38]

The widow kept her family together with modest financial help from her cousin Richard M. Hoe, the inventor and manufacturer of the lithographic rotary printing press. Between Sunday school and his mother's help, Barrows learned to read. At nine he was hired by Hoe at a dollar

---

[37]See the late Dr. Lawrence A. Frost's sadly one-sided and often factually inaccurate portrait of Stanley in *Custer's 7th Cav and the Campaign of 1873* (El Segundo, Calif.: Upton, 1985).

[38]Isabel C. Barrows *A Sunny Life: The Biography of Samuel June Barrows* (Boston: Little, Brown, 1913), 2–3, and *The Dictionary of American Biography*.

per week to run errands, keep files, and so on. Adding to Barrows's difficulties were his weak lungs and being effectively blind in one eye after being hit by a rock during a street fight. Barrows worked six days a week and, being brought up in a strict Baptist regimen, attended multiple sessions of church on Sundays.

During his teenage years, Barrows taught himself telegraphy, telegraphic equipment repair, shorthand, classical Greek, Latin, Hebrew, Spanish, French, and German. He utilized his telegraphic skills all his life; during dull dinner parties he would quietly tap on his glass in Morse code and, if getting a response, would communicate with his newfound friend—one of whom turned out to be a pretty girl. After the Civil War broke out, the six foot tall Barrows attempted to enlist in the navy but was immediately rejected because of his "enfeebled health."[39]

That year Barrows left Hoe, becoming a six dollar a week stenographer. However, his health soon collapsed, forcing him to enter a water-cure sanitarium in Dansville, New York, in the west Finger Lake region. Payment was made by his becoming secretary to the director. Not until 1873 did his health turn better—the result, he always felt, of the rigors of outdoor living.

In 1865 he met Isabel Hayes Chapin, whose husband, a missionary, had died earlier that year in India. Trained as a nurse, Isabel was working at the sanitarium. In June 1867 the two were married in Brooklyn by Henry Ward Beecher, one of the era's leading ministers and a friend of Isabel's family.

Barrows, always pencil thin, now became a stenographic reporter specializing in criminal trials for the *New York Tribune* and *New York World*. Coming to the attention of Secretary of State William H. Seward in 1867, he was invited to become Seward's personal secretary at $1,600 a year. Barrows found the work fascinating: on being summoned by the tinkling of Seward's small bell, he would transcribe Seward's letters, dispatches, treaty drafts, and so on. As far as is known, he continued into Grant's administration, serving Secretary of State Hamilton Fish. During this period Isabel completed her studies in ophthalmology, becoming one of the nation's first female physicians. However, the age's prejudices saw her unable to develop a private practice and only able to find part-time work at the Freedman's Hospital in Washington, D.C.

---

[39] Barrows, *A Sunny Life*, 54–55.

In 1871 Barrows, age twenty-six, decided to enter the ministry as a Unitarian. He resigned from the State Department and enrolled at Harvard Divinity School, his previous formal education consisting of exactly ten months in public grade school. At Harvard, he earned his way by becoming secretary to the famed Harvard geologist Louis Agassiz. Needing more money, in 1873 he accepted the summer position of the *New York Tribune's* correspondent for the Yellowstone survey.

While there are many of the day's ruffles and flourishes in Barrows's reports, his ability to take shorthand gives his quotations an authenticity that few 1873 documents capture. His approach was that of the enlightened gentleman observer; he exhibited limited emotional involvement but had an immense fascination for the geography, geology, natural history, people, and food he encountered. Barrows had the rare ability to make many of the expedition's figures three-dimensional while always possessing his wry sense of humor. While Barrows seldom wrote two words when he could use three, nevertheless even impatient readers will find themselves fully rewarded.

Barrows impressed all he met, easily mingling with officers, engineers, and Scientific Corps members. In addition, he and Custer got along famously, Barrows being the only correspondent with Custer during the chase of Sitting Bull. He then joined Custer on the 1874 Black Hills Expedition. Two years later, the *Tribune*, at Custer's request, again asked Barrows to accompany Custer during the Indian campaign. Fortunately, Barrows had just returned from Germany and felt the necessity of finding a ministry for himself that summer.

Like Custer, Barrows was a teetotaler. He had no patience with alcoholics, and after Stanley began drinking heavily, Barrows almost ceased mentioning him. Profoundly sympathetic to the plight of the Indians, he reflected the day's progressive feeling that, for them, assimilation was the correct and only Christian answer to their survival. He and Isabel adopted an Indian boy, although the child later succumbed to tuberculosis. Later in his life Barrows strongly supported African American and women's rights, offended many during his one term as congressman, and played a major role in the penal reform movement.

Barrows's reports to the *Tribune* are complemented in this book by the diaries and correspondence of military and civilian participants. These included Lt. Col. Luther P. Bradley and Edward M. Konopicky, age thirty-two. Konopicky had studied at Harvard, and before returning to Europe, he jumped at the opportunity to join the Scientific Corps as its artist. Surveyor Montgomery Meigs was the talented son of the army's quartermaster general of the same name. William F. Phelps, fifty-one, was one of the leaders of the normal school (teachers' colleges) movement in the United States and an "observer" for Minnesota governor Horace Austin. Playing a key role was the Northern Pacific's lead surveyor, Thomas L. Rosser, whose letters and diaries are also reprinted. Additional materials from other individuals (including Phil Sheridan) and various newspaper articles are included.

Custer and the 1873 Yellowstone Survey tells the story of the 1873 survey chronologically. I hope that you, the reader, will be as fascinated as I was reading these often surprising documents for the first time.

PART I

Preparations

CHAPTER I

Through June 12

The expedition of 1872 ended not only with Sitting Bull and his men turning back the western Yellowstone surveyors but also the killing of the first cousin of President Grant's wife and two others in ambushes. Thus the army, stung by these setbacks, began preparing for an 1873 campaign. A critical question was if the financially strapped Northern Pacific would continue its surveys and therefore again need protection.

This chapter begins with the initial planning and concludes with the gathering of the survey's participants at Fort Rice on the Missouri River. Reaching Fort Rice was not without incident. The Seventh Cavalry met in Memphis, slowly came up the Mississippi and Missouri by steamer, and in southern Dakota was caught in a horrific April snowstorm.[1] Three of the five Scientific Corps came from Harvard, including Austrian artist Edward Konopicky, a man fascinated by everything he saw in America. Lt. Col. Luther Bradley arrived on a smallpox-infected steamer carrying hundreds of soldiers. One surveyor, Montgomery Meigs, age twenty-five, single-handedly rescued the riverboat *Ida Stockdale* from destruction. Ex-Confederate general Tom Rosser, who'd barely survived an Indian ambush the previous October, was torn between his wife's fears and his love of adventure. Four newspaper articles describe the arrival of the soldiers, the completion of the railroad to the Missouri, and the wide-open, newly named frontier town of Bismarck.

---

[1] Roger Darling, *Custer's Seventh Cavalry Comes to Dakota* (1989) is entirely devoted to the "Dakota transfer."

The chapter's first letters are from Lt. Gen. Philip H. Sheridan[2] to Northern Pacific president George W. Cass.[3] Perhaps the most significant part of the letters is Sheridan's bowing to the Northern Pacific's request to survey the western (northern) side of the Yellowstone, not the seemingly easier-to-build eastern shore.[4]

In late 1870, after a long career constructing bridges, canals, and railroads, W. Milnor Roberts (1810–81) was appointed chief engineer of the Northern Pacific. A pacifist Quaker and gentle man (although ironically he had Union and Confederate generals as brothers-in-law), Roberts was a longtime professional associate and personal friend of Jay Cooke. Roberts was calm, logical, honest, and amazingly energetic. He had a light sense of humor, exhibited common sense, and was able to encourage and get the most from those who worked for him. Although Roberts had been ignored under the previous railway president, Gregory Smith, Cass treated him as a close advisor. Roberts moved easily among military personnel; a daughter, Annie (1849–1914), was married to George W. Yates (1836–76) of the Seventh Cavalry. While these letters do not reflect it, Roberts formed a solid friendship with Custer. His youngest son later became Custer's secretary and narrowly missed being with him at the Little Big Horn.

Monty Meigs (1847–1931) was perhaps the best trained of any of the engineers, his formal education including two years of advanced study in Germany. Meigs is often confused with his father (of the same name), who in 1873 was in his second decade as the army's quartermaster general. The older Meigs (1816–1892), an outstanding architect and civil engineer, is perhaps best known for the basic design of Arlington Cemetery, where his son John is buried. In one of the war's tragic ironies, John Meigs was killed or murdered—the issue was never really settled—by Confederate irregulars who nominally reported to Rosser, thereby causing the type of love-hate feelings that Meigs felt toward Rosser that

---

[2]In 1873 Lt. Gen. Philip H. Sheridan (1831–88) commanded the Division of the Missouri, the army's largest field command, which stretched from Texas to the Canadian border with over half the Regular Army in his jurisdiction.

[3]George W. Cass (1810–88), West Point '32, joined the Army Corps of Engineers and had a long railroad career. He also played a major role in building the National Road (turnpike).

[4]The Northern Pacific motive was surprisingly altruistic; treaties with the Crow nation had given them the eastern bank. Although the NP did everything possible to respect the treaty, by 1880 it had determined to build on the eastern side, but then paying the Crow a token sum for hundreds of square miles.

psychologists thrive on. Meigs's letters to his parents and a sister have never been published.

Thomas L. Rosser (1836–1910), who missed graduating from West Point by weeks when he resigned to join the Confederate army, became a protégé of J. E. B. Stuart and rose to a major general's two-star rank. At West Point, he and Custer lived on the same floor for four years, becoming close friends. In 1864–65 they fought each other in at least a half-dozen pitched battles. When Custer completed Lee's encirclement at Appomattox, Rosser's cavalry was the last to break out—Rosser not letting himself be captured until May.[5]

After the war Rosser struggled to find employment. Not until January 1870 was he hired by the Northern Pacific, rising rapidly in the Surveyor Corps as his leadership skills quickly became apparent. In 1871 Rosser was the obvious choice to lead the railroad's engineering team into the Yellowstone valley. In 1872 he again led the surveyors, but this time he barely survived the Hunkpapa ambush led by Gall that killed Lt. Lewis Dent Adair. Like all other surveyors, he was laid off in late 1872 but was quickly reappointed by Roberts when the Northern Pacific and Cooke scraped together funds for 1873.

Although Rosser was a large (6'3" tall, 230 pounds) and powerful man, the reader will note his frequent illnesses (as do his 1871 and '72 diaries) and can determine for themselves if Rosser was a mild hypochondriac suffering frequent depressions. Although Rosser left the survey after it reached the Yellowstone in July, the Solomon-like role he played in the Stanley-Custer relationship could not have been more important. Rosser's field diary and letters to his wife are included here.[6]

Lt. Col. Luther P. Bradley (1822–1910), from New Haven, a handsome and fashionable dresser, was the oldest and third-ranking officer (behind Stanley and Custer) on the 1873 survey. While he had not attended West Point, Bradley, a bookkeeper by profession, had been active in the Connecticut and Illinois state militias. He consistently saw combat during the Civil War, served with Stanley in the Army of the Cumberland, and was seriously wounded twice, emerging as a brevetted brigadier general. Bradley's 1873 diaries and letters to his wife have never been published. While most of that summer's letters to his wife survive, the

---

[5]For Rosser, a significant difference existed between capture and surrender.
[6]Previously published by the editor as "Thomas L. Rosser and the Yellowstone Surveying Expedition of 1873," *North Dakota History* 70, no. 3 (2003).

two sent when the survey was camped on the Yellowstone and Stanley was drinking heavily are missing. A coincidence?

Edward M. Konopicky (1841–1904), from Austria, a naturalist painter, jumped at the opportunity to join the Scientific Corps as its artist. His detailed letters, sent to his parents in Vienna and previously unpublished, are written with his European perspective and offer a candid if often unintentionally humorous account of the survey. A classic "tenderfoot," Konopicky was barely able to ride, tired easily, and was the subject of numerous practical jokes. Sadly, none of Konopicky's 1873 drawings survive, his sketch books having been destroyed during the 1945–46 Russian occupation of Vienna.

The reader should note that all references to mail sent or received, repetitious comments in diaries and letters, and discussions about subjects and individuals not germane to the context of this book have been omitted.

Philip H. Sheridan to George W. Cass

Headquarters Military Division of the Missouri
Chicago
January 27, 1873

George M. Cass, Esq.
President, Northern Pacific RR

Sir,

I respectfully request of you as the President of the Northern Pacific Railroad information as to the intentions of the Company in reference to the progress of the road west of the Missouri River during the year 1873.

I do this that we may be able to meet the wants of the Company by the concentration of the necessary number of troops to give adequate protection and in order to be able to get authority from Congress and the construction of one or more military posts between the crossing of the Missouri and the mouth of Powder River on the line of the railroad.

We now have under consideration the changes in troops and the expenses which our duties for the coming spring and summer will bring upon us; and an early reply to this letter is especially requested.

Source: Northern Pacific Records, Secretary's Unregistered Letters, at the Minnesota Historical Society, St. Paul.

Feb. 24, 1873

Dear Sir:

I have the honor to acknowledge the receipt of your letter of Feb. 5th 1873. I was in Washington when it reached my headquarters but the substance of it having been sent me by telegraph I immediately asked the Secretary of War to increase the appropriation for barracks and quarters to the extent of $250,000 for the necessary shelter of troops engaged upon your line of railroad west of the Missouri River.

I had previously, in contemplation of your road being pushed west of the Missouri River, succeeded in getting the 7th Cavalry transferred to the Department of Dakota. I also intend to transfer an additional regiment of infantry up the Missouri River.[7]

---

[7] Most 1873 regiments consisted of twelve companies, each with forty-five to fifty-five officers and men.

This force, with troops which may be made available on the line of the river, will furnish an escort of at least 2,000 men for the protection of your surveying and working parties during the coming summer. Of these 2,000 men, 1,000 will be mounted and [include] one of the best regiments in the service.

This I think will be sufficient for your operations between the Missouri and the mouth of Powder River.

I wish to state that in conversation with Genl. Rosser he thought it would be best to push the surveys from this [western] side & connect with the surveys made from Fort Ellis [Montana] down the Yellowstone River and up the Musselshell. This would be exceedingly advantageous to us and there would be no trouble except in crossing the Yellowstone. To overcome this difficulty I wish to state that we have [*unclear*] five ferry boats at Fort Buford [northern Dakota] which can be sent up the Yellowstone at least a distance of 70 miles to the foot of the rapids and by a slight deflection of the command to that point it could be crossed and the surveys continued on the west side until they met with those already made.

I also wish to say that plans and specifications of boats which can be constructed on the Yellowstone near Fort Ellis have been sent to these headquarters by Lieutenant Doane[8] and that it is more than probable that I shall order [*unclear*] of these boats constructed and bring them down to the upper end of the same rapids so that the crossing can be made either above or below.

The great trouble and expense of transferring additional troops to Fort Ellis in Montana makes me exceedingly anxious that there should be only the surveying parties from this side. I have also made arrangements to load a boat for Buford as soon as the ice goes out of the river which is to be the very first boat up of the season, and at Buford will put a strong guard on her and send her up the Yellowstone as far as the rapids and if it is possible for her to pass the rapids she will ascend as far as Powder River.

I submit this information for your consideration and I deem it very necessary that you should keep me posted. All of your letters will be considered strictly confidential should you so choose, but we want early information of your movements. It is my intention to build a large post

---

[8]Gustavus C. Doane (1840–92) was known for his participation in the early Yellowstone Park explorations. His plan was never executed, steamboats being used instead.

where the railroad crosses the Yellowstone and another, but smaller one, where it crosses the Little Missouri, but no work on these posts can be commenced until the points of crossing by the railroad are absolutely fixed.

Source: Montana Historical Society, Helena, folder SC-1889.

W. Milnor Roberts to Cass

Engineers Office
23 Fifth Ave, NYC
Feb. 5, 1873

Sir:
    Referring to further surveys, etc. west of the Missouri river:
    The distance in round numbers, from Fort Abraham Lincoln at the proposed railroad crossing of the Missouri [Bismarck] to the mouth of Powder River is 270 miles; thence to Fort Ellis, [another] 350 miles by any practicable wagon route.
    Between Fort A. Lincoln and the mouth of Powder River troops and wagons can travel at all times without delay on account of streams; the Little Missouri being the only considerable stream which it is necessary to cross; whereas with a wagon train between Fort Ellis and the mouth of Powder river it is necessary to cross the Yellowstone a number of times, and perhaps also the Big Horn, Powder river and other streams.
    The Yellowstone is sometimes not fordable till late in the summer, and throughout the season it is difficult along considerable stretches to find practicable fords, so that without the aid of boats and appliances for crossing a large body of troops, it would be difficult to furnish effective escort for a party surveying the railroad line in the immediate valley of the Yellowstone in the face of any considerable number of hostile Indians. . . .[9]
    Fort Abraham Lincoln is therefore much the more advantageous point *as a base of supplies for a Fort at the mouth of Powder River.* [Roberts's emphasis]
    If the Powder River should be considered too far from the Missouri for the first Fort west of Fort A. Lincoln, then the Little Missouri offers the most convenient intermediate point, about 160 miles west of Fort

---

[9] A discussion of the pros and cons of different supply routes is omitted.

A. Lincoln, for the protection of engineers and workmen on the line between the Missouri and the mouth of Powder River. . . .

West of the Powder River we have about 150 miles of unsurveyed ground. Should it be desired to determine the route between the Powder River and the country in Montana west of the Belt Range, *this year* it would be necessary to have a party protected by a very strong escort to make the requisite examination—to ascertain how far up the Yellowstone it may be advisable or necessary to continue in the Yellowstone Valley; whether all the way up the Yellowstone, or leaving it to pass into the valley of the Musselshell.

If there should be a Fort established at the Little Missouri, or at Powder river or at Tongue river, some sixty miles farther up the Yellowstone, any of these points would be a better base or starting place for that particular survey than Fort Ellis would be.

Source: Northern Pacific Records, Secretary's Unregistered Letters, Minnesota Historical Society, St. Paul.

Roberts to Thomas L. Rosser and Cass

Bismarck, D.T.
June 2, 1873
Gen. Thomas L. Rosser, [Chief] Engineer, Dakota Division

Your previous surveys west of the Missouri River as far as the Yellowstone have established the general features of the region; it is now proposed [for you] to survey and locate another line by a more direct route passing north of the former lines, as suggested by yourself, based on the topographical knowledge you have acquired. You will of course apply your best judgment in the exploration and location of this northerly line.

The primary object of the season's campaign is however to survey the Valley of the Yellowstone from the point or points to which your former surveys reached to the eastern termination of our surveys down the Yellowstone by Mr. J. A. Haydon, Assistant Engineer.[10] This will give us a continuous surveyed route between Lake Superior and Puget Sound.

---

[10]John A. Haydon (b. 1827), a wartime Confederate engineer, led the Northern Pacific's western survey. He was not related to Yellowstone Park explorer Ferdinand Y. Hayden (1829–87), with whom he is sometimes confused.

Mr. Haydon's instructions were to survey down as far as the mouth of the Powder River; but untoward circumstances changed his program, and the surveys stopped about a hundred and fifty miles (more or less) short of Powder River. He was also instructed to survey a line from some common point on the Yellowstone passing over the dividing range to the valley of the Musselshell, and hence westward up that valley. Owing to the trouble with the Indians the survey down the Yellowstone was abruptly ended. Mr. Haydon then passed with his engineering party, escorted by Col. Baker, across the dividing ground into the valley of the Musselshell, but without *surveying* the route across.

The second object of the present beyond the western point of your surveys of 1871 and 1872, and also as a continuation of the line you will run this summer from the Missouri to the Yellowstone, is to ascertain the most practicable and most advantageous route for passing from the Valley of the Yellowstone to the Valley of the Musselshell. . . .

In my interviews with Lt. General Sheridan on this subject in both New York and Chicago, he expressed an opinion in favor of carrying the surveys on the south side of the Yellowstone first on account of the greater value of the land on that side, and secondly on account of the importance of making such a thorough survey of the region that there would be no necessity for another costly expedition like the present one.

I stated to General Sheridan that I had contemplated moving the line of survey along the north side of the Valley of the Yellowstone, mainly on account of the presumed trouble of crossing Powder, Tongue, Rosebud and the Big Horn, all of which enter on the south side; and because our survey, after making a junction with Mr. Haydon's line of 1872, is to pass out of the Yellowstone Valley on its north side and over to the Musselshell. . . .

General Sheridan remarked that the expedition would of course go wherever we designed our surveys to be made; that is to say, on either side of the Yellowstone; and that the troops could cross it, as necessary, above Powder River; but that he would have a steamer with supplies, at some convenient point below Powder River, which steamer would also serve to cross the troops and our engineering party to the north side of the Yellowstone; and that when the expedition should return it would also serve to cross them to the south side. . . .

Yourself and party are so familiar with the requirements and duties of the contemplated surveys as to render unnecessary any detailed instruction respecting the method of conducting them. It is desirable, if occasion offers, that you should send reports from time to time of the general features of the country, with such news as may seem to you pertinent, addressed to me at headquarters in New York. . . .

June 3, 1873

Genl. Geo. W. Cass,

Our bridge-building and track laying parties have worked faithfully and energetically ever since I have been out here to push through to the Missouri river, and I have no doubt that they have been doing so all the time; and but for the serious interruptions and delays caused by frequent and heavy rains, the track would have been laid much sooner. It has rained during the greater part of 14 days out of 31, since the resumption of work. Apple Creek, which is very crooked as you saw, is crossed seven times within a few miles, and the materials for these pile bridges had to be brought out on the track. These bridges are from 80 to 100 feet long. We crossed the last one with the train this morning at seven o'clock. The rains have made the hauling very heavy; in some places requiring double teaming to get the cross ties delivered.

The track will reach Bismarck this evening. . . . General Rosser will go back this afternoon to Fargo and Minneapolis to complete his arrangements for starting with the expedition. From the tenor of telegrams he has received from his wife I think that his decision to go even part of the way will be very painful.

From the confidence imposed in Mr. Eckelson[11] by Genl. Rosser, from the fact of his having been two seasons Gen. R's Chief Assistant on the two previous surveys to the Yellowstone; from what Mr. Linsley[12] has told me about him; and from what I know of him personally, I am not afraid to trust to his skill and judgment in conducting the surveys along the Yellowstone, and on to the Musselshell and back to the Yellowstone.

. . . If I can get an escort from Fort Abraham Lincoln, I propose spending a few days on the west side of the river [myself] in arranging the line for the first eleven or twelve miles, to the point where it will pass out from the main valley of [the] Heart River.

---

[11]Albert O. "Eck" Eckelson (ca. 1845–1902) was a self-taught engineer of considerable ability and popularity.

[12]Assistant chief engineer Daniel Linsley (1827–89), a protégé of ex-president Gregory Smith, resigned in late 1872 before Cass could fire him.

General Terry[13] and staff are to start on [June 8] from St. Paul. We will arrange to take them down to Fort Rice, and tent them there, as Fort Rice is very crowded. . . .

It may easily happen that even if the Musselshell route should be shorter, the Yellowstone route might still be preferred, on account of the greater extent of the Valley, and its value for local business, coal, etc. . . .

Bismarck
June 5, 1873

Dear Sir:

Our track layers performed admirably on June 3rd and laid the track fairly into the town. They laid 3-79/100 miles, from 7 A.M. till 10:30 P.M., and had some [minor] delays. This I think is the greatest length of track laying in a day, with any of the regular gang, in my experience.

The people, including some fashionably attired ladies [of the night], were out to meet them during the day, encouraging the men, doubtless to obtain the goal. As might be imagined Bismarck was lively during the night, and yesterday morning the killed and wounded far outnumbered the live working men; but they were only "killed" by the whiskey. Mr. Winston[14] and his men tried in vain to rally a corporal's guard in the morning to resume work. In the afternoon he got out a remnant and laid half a mile. This morning he has a respectable force, and [*unclear*] says they will lay the end of the track today.

A telegram from Mr. Mead[15] says that freight will arrive here on the 10th, Tuesday next. Everything will be ready in the way of sidings . . . and that General Terry and party will be here on the 11th.

The river raised about six feet after you left; it is now slowly falling.

The steamer *Key West* is expected up the river daily with supplies for the Yellowstone. She is to be held here till one of Coulson's steamers with supplies, etc. comes up, then both are to proceed according to such orders as General Terry may give.

Gen. Rosser went back on the evening of the 3rd to make preparations

---

[13]Brig. Gen. Alfred H. Terry (1827–90) commanded the Department of Dakota (Minnesota, Dakota, and Montana) and reported to Sheridan.

[14]Philip B. Winston (1845–1901) was one of Rosser's brothers-in-law.

[15]G. W. Mead, about thirty-five, was general manager of Northern Pacific Railroad operations in Minnesota and Dakota.

for starting with the expedition. The last telegram from his wife was in effect that she could not be reconciled to the idea of his crossing the Missouri River. It has been rumored that Indians have said "if they could kill big General Rosser they could stop the railroad." Whether any such rumor has reached Mrs. Rosser I know not. Very likely it originated with some white Ass. They are plentiful at the front.

I have nearly finished the examinations I wish to make on this side of the river, and I have shown Mr. Meigs on the ground what lines I want run.

I have not yet been to Fort Abraham Lincoln or Fort Rice. Hope to visit both soon. . . .

Source: Northern Pacific Records, Secretary's Unregistered Letters, Minnesota Historical Society, St. Paul.

Montgomery Meigs to His Family

St. Paul
April 6

I am to start up the road tomorrow . . . and as Gen. Rosser goes up on Monday I shall go to Fargo with him and shall prepare to cross over to the Missouri River.

The weather is very mild here and all the streets afloat with mud after the good old style of Washington during the war. It is astonishing to see how many people are out of employment. Every other man one meets seems to be "loafing" as they term it, and applications for work on the engineer force of the N.P.R.R. pour in incessantly. Gen. Rosser has already about a thousand letters asking for positions all the way from Resident Engineer to axeman, and is besieged with persons anxious to get the work. About half of this northern population seems to hibernate during the six winter months and emerge in the spring like other hibernating animals hungry and poor after their long inactivity.

Real estate investments seem very profitable here and I understand from business men, that 25–30% profit yearly is about what is considered the usual advance. Our lots [in Bismarck] are going to be worth something. Gen. Rosser was out there a short time ago and reports great activity. . . . He sold two lots while he was there for $900 cash which he gave $250 for last summer. My best lot is on the main street and one of the best in town and I hold it at $500. The R.R. being diverted into the town will make a great difference in the sale of the lots and I expect to see a flourishing colony there before long.

Eckelson is here, and [H. R.] Stevens, Assistant Engineer, who was with me last summer, is going out again and will complete his part of the work unfinished when the track stopped last November. In Brainerd the N.P. offices seem in somewhat of turmoil. I think Mr. Meade is going in for a new deal all round and dismissals and resignations are the order of the day. . . .

Fred Grant's[16] appointment to Gen. Sheridan's staff seems to be making a great noise and creating a great deal of disgust. . . .

Edwinton [Bismarck]
April 23, 1873

We had a toilsome trip of it out here. The snow had melted and all the sloughs were full. In some places we had to wade swift streams and at others we were obliged to shovel a way through drifts of snow sometimes as much as seven or eight feet deep. Our progress was consequently quite slow. At the Sheyanne [Valley City, North Dakota] we found a rapid and deep stream and as the road bridge was swept away I got a flat car, and in three trips brought horses and all over the railroad bridge. We used to run the car from the west to the east side empty and up a steep grade, where after loading it would run down and across the bridge by its own weight. The day after we got across there came up a terrible snow storm which kept us in camp three days in all.

At Jamestown we got some more provisions and proceeded without further difficulty excepting one deep stream we had to ford with all our baggage supported above the wagon boxes into which the water poured. I am now running a new line here and the [track's] graders are expected every day.

The weather is very disagreeable and cold. A north-east wind has been blowing for three days and it penetrates to the bone. We are all

---

[16]Frederick D. Grant (1850–1912), West Point '71, had been appointed a lieutenant colonel on Sheridan's staff. Congressional authorizations allowed only so many officers of each rank; officers with lower seniority were bumped with a loss of pay to allow for Grant's promotion. He was headstrong, offensive, and drank heavily while surrounded by acolytes seeking a good word to the president, including Custer.

unaccustomed to the cold weather and I suppose that is the reason we feel it so much. . . .

I find pretty much the same set of officers as when I left last fall. They have had a fearful winter here and suffered a good deal. The Bottoms along the Missouri were flooded in some cases six feet deep and many settlers were obliged to get on the roofs of their houses to prevent drowning. The R.R. Company had a steamboat here (the *Ida Stockdale*)[17] which the ice carried away and came very near crushing. It would have been lost had it not been for some men who boarded her and at the risk of their lives. . . .

Edwinton [Bismarck]
April 29, 1873

. . . I find that we just got out here in time. The country west of Red River is all under water. Moorhead on the S.E. side of the river has had three feet of water in its streets and the sheet of water extended back some ten to fifteen miles. At Fargo which was out [of the] water the Hotel[18] was crammed with refugees to the number, it is reported, of over 200 persons, no small proportion of the population of that part of the country. The trains on the Minnesota side were delayed some three or four days by the freshet but no great damage was done that I know of. Everything is quiet here though we expect fifty teams and seventy-five men tomorrow to go to work grading. . . .

End of the Track[19]
May 11, 1873

. . . I spent last Sunday on the company steamer.[20] This steamer was carried away by the ice last spring and landed thirty-five miles down the river. I was ordered to go down and take charge of her. I found her wheel broken and sent men to repair it, hired a pilot and engineer and

---

[17] The *Ida Stockdale* was a stern-wheeler of 364 tons built in 1866–67 near Pittsburgh.
[18] The Northern Pacific's 125-room Headquarters Hotel.
[19] Thirty-three miles east of Bismarck.
[20] Despite its cash problems, Northern Pacific president Gregory Smith purchased the *Ida Stockdale* in 1872.

started for Edwinton. We got onto an infernal bar in the river and laid there for two days. We finally got off and steamed up the river. We got into another bad place, smashed the wheel and drifted helplessly against the bank where we succeeded in securing the boat after a struggle. She lies there now and we are to make a final effort to bring her to Edwinton day after tomorrow. I played captain during the voyage but as our crew was small, I played also [*unclear*], longshoreman and engineer. . . .

End of Track[21]
May 11, 1873 [Exact date unclear; likely May 14]

. . . [E]very good day brings us from one to two and a half miles nearer. I have a car fitted up something like a "palace car" only not so handsome. The upper berths fold up in the day time, and the lower ones being arranged along the sides of the car lengthwise, we have a good deal of room. The telegraph operator has a table at one end of the "sitting" room and the kitchen is at the other end of the car leaving a sanctum between used as an office and dining room by day, and at night as Gen. Rosser's and my own dormitory. My duty at present is superintending the laying of track and putting up the bridges, as well as the construction of a new grade down to the Landing at Edwinton or "Bismarck" as they have resolved to call the place where the N.P.R.R. will cross the Missouri. There [are] about 200 men now at work between here and the river and we hope to get all finished by June 1.

I have had some quite novel experiences lately in the steamboat line and you will perhaps laugh at the idea of my commanding and steering one of the largest steamers on the Missouri, but I did it and successfully too, in a place where a licensed pilot had failed.

The Company steamer was carried off by the flood of ice which poured down the river early in the spring. I was instructed on my arrival at the Missouri to get possession of her and do what I thought best in the case. Her wheel had been smashed by the ice though she was tied up on the bank in a safe place so I sent carpenters on board to repair it. At the same time, I heard that Indians had been prowling around her and a large war party being reported on the west bank of the river where she

---

[21]Twenty-five miles from the Missouri.

lay. I sent a messenger to one of the forts near there[22] and had a guard of soldiers put on board. When her wheel was ready I took an engineer, pilot and fireman down in a small boat to where she lay forty miles below Edwinton and we started up to town.

It was moonlight the evening we started and contrary to my better judgment I yielded to the pilot who said it was perfectly safe to run on such a night and in half an hour or so had the satisfaction of being aground on a sandbar in the middle of the river. All the next day we sparred and swung about to no purpose and I completely wore my hands out pulling and hauling at Manila rope besides being covered with mud and dirt. Towards evening the pilot gave the job up and started for town saying that if we would send for him when we got off he would come down again. That afternoon a steamer passed up the river and we borrowed a cable of [his]. This we made fast to our boat. . . . It was dark before we got all this done but next morning early we put out one of the spars, gave a push and to our intense delight the steamer sailed off into deep water. A messenger was dispatched to Edwinton and at four o'clock the pilot came down, we got up steam and resumed our journey. It was again a moonlight night and the pilot wished again to try a moonlight voyage. I had had enough of that however so we tied up to the bank for the night at a safe place. Next morning we cut some wood and got up steam and all went well until we got within about ten miles of Edwinton. As we passed the sawmill we blew our whistle and brought everybody out but "pride goeth before the fall" and so it happened to us. . . .

[T]he river is divided into three channels "A", "B" and "C." Now no pilot on the Missouri can be sure of the right channel in such a place without trying them all, partly because the channels shift sometimes inside of twenty-four hours and partly because the water is so rapid and so muddy nobody can tell without sounding how deep it is. In the present case our pilot made a mistake and took channel "A" which appeared to have plenty of water but is narrow, crooked and obstructed by sandbars.

We had not gone halfway up when the steamer grounded and as her stern swung down-stream with the current the wheel struck the bank and was in an instant smashed all to pieces. The boat began then to drift helplessly down the stream. At last we neared the bank and just at the right moment we launched a boat, succeeded in landing a cable and

---

[22] The Grand River Indian Agency was twenty-five miles below today's border between North and South Dakota.

secured her from drifting further. Her wheel disabled, our voyage was at an end. So we landed all crew but carpenter, blacksmith and engineer and left them to repair damages.

Before the wheel was ready again my pilot went down to Sioux City and as there was no one left at Edwinton who knew anything about steering a steamboat, I came to the conclusion that as the boat must be brought up to town, it would be more satisfactory to myself to manage her myself even if I got her into difficulty again.

This time I took no crew down but the man in my engineering party and a steam engineer. We found the wheel not quite ready but started fires under the boiler and by two o'clock had steam up and wheel ready to start.

As I had never steered anything bigger than a sailboat before, I took my place at the wheel in the pilot house in some little doubt as to how she would act and did not feel quite sure I could get her away from the bank against which she lay. I ordered the lines cast loose, however, rang the bell for the engines to start, and a slight wind helped us so we were soon floating in mid channel. Away we went upstream sounding occasionally to keep in the deep water and I began to feel more confident.

We had not gone far before I found that there was no great amount of intellect required to manage even a large steamboat as she obeys the same laws as the smallest sailboat, only one rows with a huge engine, and steers with a wheel. We had to land and take on more wood. "All ready with the line," I shout. "Aye, aye, Sir"—jingle goes the bell, the engine stops and we come slowly up to the bank. The men jump off and fasten a cable round a large tree. I start the engines again for a moment to keep the boat from swinging downstream too rapidly and in another she is lying quietly at the wood yard. While the men are carrying the wood on board I walk into the woods and get a glass of native wine manufactured on the island. Capt. Clarke[23] who commands the post at Edwinton accompanies me and after half an hour or so we go on board again and prepare to start. I throw the wheel over, ring the bell to back the engine, cast off the lines and we back rapidly out into the stream. Again I ring and the engine stops. Another ring and we begin sailing upstream again, through channel "B" which we had sounded on our way down. Imagine my [unclear] when the steamer "stubs her toe" as it

---

[23] Capt. Charles E. Clarke (d. 1901), Seventeenth Infantry, was a Civil War veteran and a member of the 1871 Eastern Yellowstone surveying expedition.

is called and quivers from stem to stern as we strike a sandbar. I back the engines however and she swings off to my great relief and we sail along without further mishap.

I had loaded one of the cannon on board and as we pass Ft. McKean [Lincoln] we shoot it off and hoist the stars and stripes. Then one of the boys who is instructed to blow a salute with the whistle pulls so hard that we can't stop it and blow off so much steam that we can barely make headway against the current. This amuses the crowd on the bank very much and they make many kind offers of throwing us a line to tow us etc. but we throw wood into the furnace and at last tie up safely to the bank at the Edwinton Landing. "*Thus Endeth the Cruise of the Ida Stockdale.*" . . .

Source: Elizabeth R. Atwood, "Letters of Montgomery Meigs Written While Engaged in the Survey of the Northern Pacific Railroad, 1872–1873," MA thesis, University of Montana, microfilm 140, State Historical Society of Montana, Helena.

Rosser's Diary, May 22–June 12

THURSDAY, MAY 22ND. Left Fargo on the 5 A.M. train without breakfast. Rained all day and I would *not* have gone through but heard of a strike at the end of track and went through to settle it. Reached end of track 11 P.M.

MAY 23RD. Compromised with the strikers by agreeing to pay them 50 cents per day more than the contractors were paying them now. They went to work. Weather bad and progress slow.

MAY 24TH. Weather continues bad. Rained all P.M. But as we reached the slough and could work only a portion of the men. . . . Slough very bad and crossing it tedious and could not use teams or even horses to iron cars.

MAY 25TH. Weather better though it rained today. Worked the larger portion of the force today and had a rough, disagreeable time of it. Did not get over the slough but made considerable progress and will get over tomorrow early.

MAY 26TH. Crossed the slough after a hard day's work. Weather warm and mosquitoes furious. Many of the men sick from drinking the water from [the] slough which is strong[ly] alkaline.

MAY 27TH. On solid ground today. The mosquitoes increasing daily. Drove to Bismarck in P.M. and was caught in dreadful storm of rain, wind and hail. Ran into bridge camp and waited till storm passed. Spent night in little office in Bismarck. Suffering of sore throat and feeling badly.

MAY 28TH. Read a telegram from precious Lizzy reminding me that this is our [tenth] wedding anniversary but I was thinking of it when her sweet message came. God bless her. I hope the time will soon come when I can spend all my time at home with my family.

MAY 29TH. President Cass and Mr. Roberts came up today and I took them down into the river. Cass was extremely cordial and pleasant. I informed them that I could not go all the way to the Upper Yellowstone on the survey. . . .

MAY 30TH. Party returned to end of track and proceeded to Fargo. Promised to finish track laying the 3rd but fear I will not quite do it. Mr. Roberts remaining with me.

MAY 31ST. Rained very hard during the A.M. but cleared off about 7:30 and we went vigorously to work and laid 150 [*unclear*] during the day. If we could go on this way, we would not be far from the river Tuesday night.

JUNE 1ST. Rained very hard all the A.M., cleared off about noon and went to work in the P.M. and although very muddy we laid about one and a half miles of track. Have promised to get to Bismarck the 3rd instant.

JUNE 2ND. No entry.

JUNE 3RD. Rose this A.M. with the intentions of making great efforts to finish the track into town and although we got a bad start we succeeded in doing so. I left the post at 4 P.M. and ran home to see my precious wife and children.

JUNE 4TH. Left Fargo for home. Had a pleasant ride. Reached Duluth, took supper and took sleeper for St. Paul and telegraphed dear Lizzy to meet me in St. Paul.[24]

JUNE 5TH. Reached St. Paul and went to Metropolitan Hotel. Took a bath and made preparations to go to meet Lizzy but learned that she had

---

[24]With Lizzy's mother living with them in Minneapolis, Rosser desired the privacy of St. Paul's Metropolitan Hotel.

come down on the St. Paul & Pacific RR and was shopping at merchants. Took her to Metropolitan then I went to see Mead and General Terry. Then came home with Lizzy and Tom.

JUNE 6TH. Spent day in Minneapolis with Lizzy and children. Had pictures taken and fixed the pretty locket which dear Lizzy gave me as an anniversary present of our marriage....

JUNE 7TH. Spent day quietly at home....

SUNDAY, JUNE 8TH. Attended church and listened to a political sermon on the labor question....

JUNE 9TH. Downtown with Lizzy. Drove out with Morrison's team in P.M. Phil [Winston] remained with us and accompanied us....

JUNE 10TH. Downtown with Lizzy trying to get a servant, but failed. Spent evening with Mr. & Mrs. Brackett....[25]

JUNE 11TH. Got pictures of Dear Lizzy and did a little shopping....

JUNE 12TH. Had made arrangements to spend evening with Mr. Morrison, but was hurried[ly] called to Bismarck to begin the Yellowstone survey. Left with a heavy heart at 7:10 P.M. Poor sweet Lizzy and the children will miss me very much but not as much as I will them.

Source: Alderman Special Collections Library, University of Virginia, Charlottesville, #1171 d-e-f, box 2.

Luther Bradley's Diary, May 22–June 13

MAY 23RD. Got to [*unreadable, likely Sioux City, Iowa*] at 11 A.M. and found steamer *Western*[26] waiting [on the Missouri River]. Embarked companies and freight.

MAY 24TH. Col. Townsend reported about 1 A.M. with [two] companies 9th Infantry and [four] 8th Infantry loaded up during the night. Had roll

---

[25]George A. Brackett (1836–1921) supplied many of the Northern Pacific's early surveys with provisions and became a lifelong friend of Rosser.

[26]The *Western*, a stern-wheeler constructed at Pittsburgh (1870–71) with a four-hundred-ton carrying capacity. A "mountain boat," she featured a shallow draft and low profile, permitting better operations when running against the wind.

calls and found we had not lost a man since the companies left garrison. Started for Fort Rice at 5 A.M. Pleasant day and made good progress.

MAY 25TH. Fine day and had a good run. Passed the Ponca [Indian] Agency about sundown, bank lined with Indians old and young. Tied up at the Yanktonai Agency at 10 P.M. on account of the darkness. Saw a good many of the Yanktons, some spoke excellent English and said they had been to church to hear their Missouri preacher, an old acquaintance Rev. W. W. Cook. Sent my card to him by one of the Indian warriors. Gave a single half dollar to "Wercona" a pretty little girl about 12 years old.

MAY 26TH. A rainy day. Steamer very wet and uncomfortable. Passed Fort Randall[27] just after noon. Took on board Capt. Poole and Lieut. Campbell[28] of the 22nd Infantry for Fort Rice. Tied up at dark on account of the storm.

MAY 27TH. Fine weather and made a good run. A case of small pox on board, one of the cabin hands, producing quite a scare among our people. Did all we could to prevent the spread of the disease.

MAY 28TH. Pleasant weather and the steamer is making good progress. Reached the lower Russell Agency at 10 A.M. Went on shore and spent an hour with Capt. Rush, whose company is stationed here. Saw plenty of squaws and children, some of the Indians at work in the corn fields. Landed the small pox patient on an island opposite the Agency and left a man to take care of him with tent, food, medicines, etc. Reached the Upper Yanktonai Agency at 2. Met Mr. Gregory of New Haven, son of the Admiral,[29] and had a very pleasant talk. Had a violent storm at night which compelled us to tie up before dark. On the post side a [unclear] had smashed into a stateroom. Kept up steam all night and engines working to keep the boat under control.

MAY 29TH. Stormy day and high winds. Made but little progress.

MAY 30TH. Storming yet. Lost a man overboard at 3 A.M. while tied to the bank. Lowered boat and tried to reach him, but the current was too strong; and he was lost. Had hard rain and high wind all day.

---

[27]In southeast South Dakota; built in 1856 to control the Upper Missouri and the Sioux.
[28]Lafayette E. Campbell (ca. 1844–1919). Known to have kept an 1873 diary, now missing.
[29]Rear Adm. Francis H. Gregory (1780–1866) was a War of 1812 combat veteran who captured or sank a half-dozen Caribbean pirate vessels, and at eighty he came out of retirement to supervise ship construction.

MAY 31ST. Storming hard, boat as wet as a washtub. Reached Fort Sully at noon.[30] Genl. Stanley and officers came on board and spent the afternoon with us. Landed 75 tons of freight and left about 7 and ran until 9 and tied up.

JUNE 1ST. Raining still. Started at daylight. Passed Cheyenne Agency[31] at 5 A.M. Weather cleared up about 9 and had a fine day. Run all night by star light and reached Swan Lake at sunrise, had some difficulty in getting over the shoals, were delayed three or four hours.

JUNE 2ND. Fine day. Passed a large village of Blackfeet Sioux on west bank of river. Saw about 100 lodges and a thousand ponies. Stopped to [get] wood about sunset at a beautiful spot opposite Buffalo Island. Reached Grand River Agency[32] at 11 P.M. and stopped an hour. Run all night by star light.

JUNE 3RD. Good weather and made a fine run. Passed the 7th Cavalry on east bank of the river in the forenoon. Overtook the steamer *Miner*[33] escorting the Cavalry with stores and grain on board and several ladies living on her. Wives of officers of the regiment. Steamer camps alongside of them every night: rather a nice way of campaigning. Run all night again.

JUNE 4TH. Reached Fort Rice at 5 A.M. Landed a mile below the Fort and unloaded stores and baggage. Marched troops off the boat at 8 A.M. and made camp on a fine bench of land overlooking the Missouri. Found plenty of fine grass, wood and water. Named the camp "Camp Terry" and it looked very nice. The camp was quarantined by the Surgeons on account of hav[ing] been exposed to small pox. No intercourse with the Post or other camps allowed.

JUNE 5–7. [No entries.]

JUNE 8TH. Laying quietly in camp, drilling and fitting out for the Yellowstone Expedition. Gen. Stanley and staff arrived on steamer *De Smet*

---

[30]Present-day Pierre, S.D. Sully was headquarters for the Middle District of the Department of Dakota and Stanley's Twenty-second Infantry Regiment.

[31]On the western side of the Missouri where the Cheyenne River enters it.

[32]The Grand River enters the west side of the Missouri just northwest of present-day Mobridge, S.D.

[33]The *Miner*, a 299-ton stern wheeler, went into operation in 1867. It was scrapped for its parts about 1877.

today and established Headquarters near us. Weather getting very hot and mosquitoes enough to eat us up.

JUNE 9TH. Custer arrived today with the 7th Cavalry and camped on the north side of the Fort. Received first letters from home, via St. Paul. Rained hard in afternoon. Quarantine raised [when] retreat [was sounded].

JUNE 10–11. [No entry.]

JUNE 12TH. Gen. Terry arrived today from St. Paul with Green, Lorris and Hughes. Spent the evening at the Post.

JUNE 13TH. Col. Lorris inspected the Battalion in forenoon. Terribly hot. . . .

Source: *Luther P. Bradley Papers, U.S. Army Military History Institute, Carlisle, Pa.*

Bradley to His Wife

Steamer *Western*
June 2, 1873

My dear love:[34]

We expect to reach the Grand River Agency this afternoon and I must send you a letter from there as the mail goes down from the Agency to Sully in a few days. I tried to telegraph to you from Sully, but the line was down and it was so uncertain when it would be fixed that we abandoned the idea of sending a greeting to the widows that way.

We have had fearful weather since we started with the exception of one day, heavy rains and wind with as heavy thunder storms as I ever saw. Yesterday it cleared up and today is fine and warm. I hope we shall have good weather now 'till after we get into Camp. I expect to be given permission to mount all the Officers and we are going to try to get the Indians at Grand River Agency to bring up a lot of their ponies to Rice, this will make the march a good deal easier for the Officers, and they will enjoy the trip all the more. Gen. Stanley told me that it was decided on to establish the supply depot on the Yellowstone bringing freight round

---

[34]Bradley married Louise "Ione" Dewy in May 1868 in Chicago, and they had two sons.

by Steamers: this will save us the hardest part of the summers work in escorting trains to Rice and back and ensure us better supplies. . . .

Everything indicates a fine and successful expedition for us. . . . We fathers, Jordan, Burt, Townsend, and the rest, console ourselves by talking about our little ones and telling all the smart things we remember of them. Jordan says he is a regular baby himself when away from home. I wish you could take a look into our cabin, half the Officers are seated writing, some in shirt sleeves, some in stocking feet.

Camp Ferry,[35]
No. 9
June 10, 1873

. . . The 7th Cavalry came in yesterday so we are all together at last and the plateau on which we are camped looks like an old war camp. The Savants [Scientific Corps] appointed by the Secretary of War to accompany are here too and are camped adjoining us. Major Lazelle[36] of my Battalion has been selected to . . . take care of them for the summer. Among them are a geologist, mineralogist, botanist, buggist, photographer, etc. They are a nice set of gentlemen, but as green as grass. We have a couple of English noblemen too who go out with the expedition to hunt and see the country. They have their own outfit of wagons, horses, cooks, etc. and everything necessary for a good time. There will be a number of private citizens from the east and west along with us too. Sheridan[37] and Terry will be here tomorrow and we shall know then about when the expedition gets off.

The steamer[s] to take our supplies round to the Yellowstone are here and we'll start—— [*section of letter missing*]

. . . I am burned as red [as] a cherry and begin to look as I used to in the summer of '67. I sleep like a top and eat *three* regular meals without making a bit of a fuss. I have got an excellent horse, the one I told you was selected for Gen. Ord.[38] He is a solid bay with fine action. Have

---

[35]*Ferry* is not a misprint for General Terry. Rather, because of the smallpox they remained quarantined until the smallpox spent its course.
[36]Capt. Henry M. Lazelle (1832–1917), West Point '55, was also on Stanley's '72 survey.
[37]Bradley was mistaken about Sheridan.
[38]Brig. Gen. Edward O. C. Ord (1818–83), West Point '39. In 1873 Ord commanded the Department of the Platte, and Bradley reported to him. Like General Terry, Ord reported to Sheridan.

named him "Sandy." I take a long ride every day and shall enjoy the time spent in the saddle this summer more than any other.

No. 10
June 12, 1873

My dear love:
Your letter of May 26th with the papers and mosquito netting came yesterday by the regular mail up the river . . . Gen. Terry came this morning and says our wagon train will not be here 'till the 15th—it will take three or four days to get it across the river and load it. So we shall not get off 'till about the 20th. . . .

. . . We are all well, and getting along nicely, only a little tired laying in camp doing nothing but drill in the afternoon about half the time and the companies drill in skirmish every day.

We have had thunder showers every day or two, but our camp is on a gentle slope to the river and drains itself. The sun is very hot, but it is only a foretaste of what we may expect next month. I have got me a large straw hat to wear in the hot weather for I find the black hats already begin to feel heavy. . . .

Source: Luther P. Bradley Papers, U.S. Army Military History Institute, Carlisle, Pa.

Edward Konopicky to His Parents

St. Paul, Minn.
June 1, 1873

After a 12-hour steamboat and a 5-hour railroad ride, Misters [Joel] Allen,[39] [Charles] Bennett and I arrived in St. Paul yesterday evening. Our itinerary was as follows: On May 27 at about 4 p.m., I and the above companions, who were highly recommended by Dr. Steindachner,[40] set

---

[39] Idiosyncratic and authoritarian Harvard professor Joel A. Allen (1838–1921) had been chosen by the government to lead the expedition's "Scientific Corps." His assistant was Harvard professor Charles W. Bennett (1828–91). The two were particularly strong in zoology, botany, and biology.

[40] Dr. Franz Steindachner (1834–1919) was superintendent of all the Austrian imperial museums.

out from Cambridge.... [W]e left Boston at 5:30 P.M. and reached Fall River in about 2 hours. From there we took a steamboat and arrived in NYC in about 12 hours. We had the day free and spent it pleasantly.... [W]e went to see the famous Central Park, and indeed, the 10-year-old park deserves to be visited by every stranger. We took a boat ride on the largest of the five ponds, drove on to a restaurant at the end of the park where we had some refreshments, toured the park, etc.

In order to give you a quick idea of the size of this park, I will only mention that it is 2½ miles long, ¾ miles wide, has 9½ miles of roads and 28 miles of pedestrian paths. New York's reservoir in the park stores 850 million gallons of water and covers 96 acres. The entire park covers 862 acres. In the evening, since my companions are Yankees who do not drink, I went out alone to sample some beer halls, and the beer was much better than any I had consumed in Boston.

At 10 o'clock we left NYC by train and slept as well as one can on a train, because in America trains traveling longer distances are equipped with sleeping compartments that can be rented for an extra $2.50 a night. When we awoke, we were already in Pennsylvania....

At about 3 o'clock we arrived in the notable city of Pittsburgh, which we could not see at first because of the thick smoke. Only after we had passed into the smoke we discovered houses and people. Actually, Pittsburgh is nothing more than a large iron foundry. Every third or fourth house is a foundry, and therefore, all the houses, streets, people, etc. are black. We ourselves, as we were having dinner, started looking sooty after half an hour.

Early on the 30th we reached Chicago, which you know is famous for its great fire of 1871. After we settled in the large Hotel Sherman, we toured the town, which has been magnificently rebuilt since the fire. It is truly a puzzle how these hundreds of palaces ... mostly made of marble, sandstone and steel could be rebuilt so quickly. On the outskirts of the town, we enjoyed the wonderful sight of a sea, Lake Michigan. On the way back, as we were looking at some remaining fire destroyed ruins, we heard the newsboys hawking about a big fire this morning in Boston....

At 9 o'clock in the evening, we left Chicago and the next morning we had the pleasure to find ourselves in the forests of Wisconsin, which were wild, uncultivated and thinly populated with a few log cabins. At 5 in the afternoon, we passed a picturesque, big black lake. Around 6:30 P.M., we reached the Mississippi and, from the high left bank, glimpsed the very beautiful city of St. Paul. An omnibus took us to the elegant Metropolitan Hotel, which I soon left, in order to admire from the heights the Mississippi flowing in its mountain bed. As I toured the city, I was happy to hear much German spoken. I also discovered many German businesses and beer halls. . . . And the best thing is the first rate Milwaukee beer. . . .

Fort Rice Dakota Territory
June 9, 1873

. . . We, that is Misters Allen, Bennett, [William] Pywell[41] and I, left St. Paul, from where you should expect to receive my letter [written] June 2nd on the North[ern] Pacific Railroad. On the morning of the 3rd we reached the wretched settlement of Thomson, which lies in a romantically wild location. We had breakfast and continued on another train after a layover of several hours.

The trip this day was something to remember as we passed through forests, meadows and lakes—I counted no fewer than 51 large and small lakes. At the Brainerd [Minnesota] Station, we again passed the Mississippi. Near[by], I had the pleasure of seeing my first band of Indians, which consisted of about 10 people and four dogs. One of them carried a boat on his head.[42]

Towards evening, we reached the prairie and stayed overnight in the hotel [at] Fargo. The next afternoon, we set out to cross the prairie, but not in a passenger car but in a baggage car. There were 10 more people in the car, with their baggage and two young pigs. The stretch from the Prairie Hotel is essentially new and since it crosses a virtually unpopulated area, it carries little traffic. I will always remember that trip across the prairie—it was excruciating. Aside from the facts that the rail bed is poorly laid and we were tossed about as if we were in a small boat, and the hard seats were very uncomfortable, the train also stopped for 3 or 4 hours under an oppressive sun. Mosquitoes, with nothing refreshing to drink, plagued us. Since we could not endure it in the car, we climbed under the car for some shade and protection from the mosquitoes. The

---

[41]William R. Pywell (1843–86), the expedition's photographer, had worked during the Civil War for Mathew Brady and Alexander Gardner.
[42]This was likely a canoe—a word that Konopicky may not have known.

first day on the prairie, we saw prairie hens, wild animals, vultures and many other birds, and among mammals, prairie dogs and antelopes. Of buffaloes, we saw mainly bleached skulls and ribs. Buffaloes were exterminated in this area seven years ago. We did not see any human beings until we reached the prairie rest stop in the evening, where we also spent the night.

The next day was a repeat of the first day's tedious trip. We stopped for four hours exposed to the sun and mosquitoes, and this night we spent on the prairie in the railcar. To keep from being eaten alive by the mosquitoes, we had to keep a fire going in the car all night. Finally, on the third day we reached the end of the prairie and the sizable settlement of Bismarck. From here, we continued by express wagon for the 5 miles to Fort Lincoln. (The railroad goes only as far as Bismarck.) After a rugged hour-long ride, we came to the Missouri River, were ferried over, and after another half an hour uphill climb arrived at Fort Lincoln. We immediately presented ourselves to the commanding officer of the Fort,[43] from whom we received a friendly welcome and an invitation to supper. We were quartered in a room in the hospital. The next day, with an interpreter, we visited the Indian camp within the Fort (about 65 Indians) who interested me very much because of their picturesque appearance (the men were half naked). The General's wife, who had invited us the next day, told me (she speaks excellent German) that unfriendly Indians attacked the fort 4 times last year and once already this year. Three Indians were killed in the last attack. . . .

On Sunday, June 8, at 4:30 [A.M.], we left Fort Lincoln to begin our actual expedition. The caravan consisted of 10 wagons, each drawn by 6 mules, a company of soldiers and 4 Indians who rode far ahead of the caravan to act as scouts and front guards. It was amusing to see how these fellows combine their scouting duties with antelope hunting. During the trip to Fort Rice, they came back with 4 dead and 2 live antelopes.

The route to Fort Rice is through barren but interesting hilly country. We reached the relatively large Fort on the Missouri River about 4:30 P.M. The General[44] himself showed us our provisional room in the barracks. Tomorrow we move into the camp set up around the Fort. The

---

[43]Lt. Col. William P. Carlin (1829–1903), West Point '50. Carlin had been a brevetted major general during the Civil War and would have been addressed as such.

[44]Likely refers to Col. Thomas L. Crittenden (1819–93), also a brevetted major general during the war.

trip to the Yellowstone River, about 280 miles from here, will begin [between] June 14 or 21. It will consist of a column of about 300 wagons with 3,000 mules, and for protection, 1,500 infantry and cavalry.

Military Camp at Fort Rice
June 11, 1873

Yesterday we moved into the camp that had been established outside the Fort. After supper with the Commander of the Camp, we sought out the tents, which were already set up for the Scientific Co[rp]s. They were a little way from the camp and consisted of a supply tent and five sleeping tents.... Behind our tents were 6 others for the cook, servant, and guard. As far as I have heard, we will start on the 14th. We have been assigned 9 horses and 2 wagons with springs. Be happy with me, my dears, about my luck that I have the opportunity [to] see this big, beautiful America, about which I had no idea during my two years in Cambridge. It is in such a unique way, because as far as I know, I am the only non-American. If you do not hear from me for awhile, even for months, do not worry. It is only because we will be cut off from communication from Fort Rice.

Source: Austrian Museum of Natural History, Vienna. Provided by Robert Pils, who managed an exhibit and lecture, "Edward Konopicky in North America," October 2002. Translated by Charles M. Baltay, Arlington, Va.

Newspaper Articles

RIVER ITEMS, LOOKING FOR BENTON

FORT SULLY, MAY 24, 1873.—... [A]fter we left Sioux City, and all got along very well until we reached Fort Randall, then trouble commenced.

At Randall we took on three companies of soldiers to go to Fort Rice, from which place they start on the Yellowstone expedition. The *Katie P. Kountz*[45] when she left Yankton was drawing nearly four feet, and the

---

[45]William E. Lass, *Navigating the Missouri* (Norman: Arthur H. Clark, 2007), 272, 301, identifies the ship as a "slow tub" and unreliable after August 1 when the river dropped noticeably. The *Katie Kountz* was built in 1871 near Pittsburgh and burnt in 1878. She was named for the daughter of steamboat captain and Missouri River businessman William J. Kountz.

soldiers put her down in the water four feet two inches. To tell the truth, there is not four feet [in the Missouri] on average, so just realize for one moment the old tub of a boat trying to ride on the ground, sparring and puffing all day long, trying to crawl a few inches every hour. Some days we did not make ten miles. One morning we started out all right . . . but alas! [soon] we were aground . . . and at night were in sight of where we left in the morning. If we ever get to Benton then God be praised. . . .

Although the *Katie P. Kountz* is the slowest boat on the river and has many difficulties on account of being loaded so deep, still we managed to pass away the time with the little excitements that always happen on a boat. The first three or four days after the soldiers got on board, I was kept very busy selling goods, but after they were possessed of their wants, they settled themselves down to their usual routine of life—that is gambling. I go down every day to watch them, and you would be astonished to see the amount of money that changes hands. Every game under the sun they play . . . [and o]ne of the soldiers won $200 in one day playing cards of some time. They had just been paid off, but as is the case generally, it gets in one or two hands at the end.

. . . The wood all along the river this year is very scarce, and we have been compelled several times to stop and cut enough to get us where they keep it. One day we stopped and commenced cutting some old wood along the bank, when a number of Indians came down and were very angry. We had to give them some flour and bacon to satisfy them. The wood was of no account to the Indians, but they thought we had no right to it. We are not far enough up the river to see game, but after we get above Rice (if we ever do) they say it is to be found in abundance. . . .

Source: Sioux City Daily Journal, *June 8, 1873. The writer is not identified.*

A SOLDIER DROWNED IN DAKOTA; LOST FOR WANT OF SAFEGUARDS

STEAMBOAT *WESTERN*, OFF FORT SULLY, D.T., MAY 30, 1873.—A sad accident occurred about twelve miles below old St. George[46] at half past three o'clock A.M. today. Private Hoffman, Company E, Ninth Infantry, fell overboard and was lost. A boat was lowered and search made, but no

---

[46]Unable to identify location.

trace of the man could be found. There were no [rope] guards whatever on the lower deck. The officers on the boat say, "Every time we carry troops up some of them get lost." If they had rope netting, or even a single rope guard along the sides of the boat, such accidents could not occur.

Source: New York Herald, June 16, 1873.

MARK KELLOGG: LETTER FROM BISMARCK;
NPRR CORRESPONDENCE—BISMARCK WEATHER—
INDIANS—FALSE RUMORS—A VISIT TO
THE "BIG INGUNS," ETC.

BISMARCK, MAY 31, 1873.—It is now three weeks since I arrived at this goal of the Northern Pacific Railroad. . . . Some facetious cuss, not having the fear of his early teaching in sight, has made a remark "that the climate of Dakota was too dry for successful agricultural purposes—that it never rains in Dakota. That chap . . . should be subjected to a Noachian [sic] deluge; for with two honorable exceptions, it has rained almost continuous[ly] for two weeks, and such rain and such thunder and lightning; and such rains as have prevailed are of the first-class order, hard to equal, never excelled in any country.

Bismarck grows fast and will assume town-like proportions very rapidly. The population comprises nearly every nationality and every known social element of the country—rollicking blades, sports, soiled doves, spirituous venders, semi-respectable, respectable, good, bad, and indifferent coming in daily contact and common intercourse. . . . There has been no quarreling, no scenes, no sensations as yet. Peace and harmony prevail, and it is hoped that this state of things will continue; but it is uncertain—the volcano exists, and an eruption, followed by a tragedy, will be a startling fact, one of these days.

. . . [The Sioux] have not been seen on the east side of the Missouri River near here or at all as far as I can learn. . . . The reports of their attacks upon Fort Lincoln, or upon Bismarck, are entirely without foundation so far. I will keep the *Pioneer* posted on matters occurring. . . .

Steamers have been arriving quite plentifully. I observe they are well laden with passengers and freight. Those going through to Fort Benton have, thus far, been carrying full cargoes. . . .

The "end of the track" was twelve miles east of Bismarck, Friday night. It will reach here next Thursday, if the weather permits. The arrival of the track is impatiently waited for by those settled here, and awaiting supplies of all kinds. The railroad will not carry freight west of Jamestown, which is 110 miles east of Bismarck, making a long tedious haul by teams, very expensive and unsatisfactory.

... "Shacks" spring up like mushrooms, in a night. A "Shack" is a sort of log or pole house, made hurriedly and for immediate use. They are queer looking habitations, generally about twelve or fourteen in dimensions, having a low flattish roof, made of the same material as the sides, and mudded over several inches in depth, to keep out the rains. ...

Source: St. Paul Daily Pioneer, June 6, 1873. Mark Kellogg (1833–76) was the correspondent who rode once too often with Custer.

MARK KELLOGG: FROM THE MISSOURI, LETTER FROM BISMARCK: SOMETHING ABOUT THE TOWN AND ITS TRADE—YELLOWSTONE EXPEDITION— INDIAN SKIRMISH, ETC.

BISMARCK, D.T., JUNE 5, 1873.—Bismarck, the present terminus of the Northern Pacific Railroad, is 197 miles from Fargo. Trains now run through from Fargo to Bismarck without laying over at Jamestown. The arrangement went into effect last week, leaving the former place at 6 A.M., and arriving here at 8 P.M. The road is in good condition, considering that it is not yet ballasted. Gangs of laborers are now at work, and in due time will be ballasted. Fares are $10 over the Dakota Division. . . .[47]

There are now in the town [500 residents,] four stores that have general merchandise, one drug store, two jewelers, one news depot and book store, one boot and shoe store, one tailor's shop, one milliner and dress maker, one gun shop, one furniture store, three hotels, three billiard rooms, one dance house, and eighteen saloons. . . .[48]

A large influx of the sporting fraternity came in early this spring, who expected that business in their line would be lively, but thus far they have had very little to do. . . .

Today a party of Sioux have been skirmishing with the Rees back at

---
[47] The rest of this paragraph concerns lumber requirements.
[48] Not included were houses of "soiled doves."

Fort Lincoln. The soldiers have shelled them from the fort, but [to] what effect is not known. One company of infantry is stationed in the town and the citizens are not apprehensive of any attack from the Indians.

Religious services have been held in Chummey's big tent.[49] A meeting will be held next Tuesday evening to take the necessary steps towards erecting a church. Hereafter religious services will be held every Sunday. Today a baseball match was played between the soldiers and citizens, the latter coming off victorious.

Source: St. Paul Daily Pioneer, *June 19, 1873.*

STRUGGLES FOR THE YELLOWSTONE

JUNE 7, 1873.—[S]ome 1,500 to 2,000 men have flocked to Fort Rice in the vain hope of being able to join the Yellowstone expedition. This the Government will not permit, and the result is that a large number are collected there who are without money [after losing everything drinking, gambling, or whoring] and the necessities of life, and having to depend entirely on the charity of the military authorities. The officers are very much annoyed by the presence of this large army of hangers on, who it has to feed and protect, and have finally issued an order prohibiting the landing of civilians at Fort Rice. When the *Far West*, on a recent down trip, sought to put off there a number of men who got on at [Bismarck] with the intention of following the expedition, she was not allowed to do so, and the result was that the boat had to bring through a lot of penniless adventurers whom they could not put ashore in the Indian Country.

Source: Sioux City Daily Journal, *June 8, 1873.*

FROM FORT RICE: YELLOWSTONE EXPEDITION ITEMS— LIVELY TIMES AT THE RENDEZVOUS— THE HOSTILE INDIANS

FORT RICE, D.T., JUNE 12, 1873.— . . . I hasten to give you the following items of news.

---

[49]Chummey, first name not known, was a gambler who had set up a 100 × 25 foot tent and adjacent "pig farm," the latter at the bottom of the prostitution scale.

... Perhaps there has been no time in history of Fort Rice when times were so lively as present. Steamboats are rushing around in the river; cavalry and infantry camps loom up in every direction; wagon trains and herds of stock are moving everywhere, and would-be contractors fly about on the wings of the wind. It's very lively indeed,

The air is full of Indian rumors. It is reported on good authority that the Indians intend to give General Stanley his first trouble at Heart's Butte, a few miles from this place.[50] This report is strengthened by the fact that nearly every fighting Indian has left the several reservations on the river, and gone in that direction. Rifles, cartridges, powder and balls are in great demand among the Indians. They will give up the last thing on earth for them . . . especially daughters. A drink of whiskey would buy them any day. . . .

Source: Sioux City Daily Journal, *June 29, 1873.*

---

[50]Actually closer to sixty miles.

CHAPTER 2

June 13 through June 17

By mid-June almost all of the expedition's personnel had reached Fort Rice. Chartered steamers were unloading thousands of tons of grain for 2,300 horses and mules, 700 head of beef cattle (to be slaughtered along the way), and arms (including two 3" Rodman rifled artillery pieces) and ammunition.[1] They brought food, clothing, tents, blankets, and personal items for 1,575 officers and enlisted men. Also unpacked were stores for the sutlers, equipment for temporary bridges, six forges with requisite horse-shoeing materials, coke for furnaces, and spare wagon parts—among the many hundreds of items necessary for a protracted campaign. What hadn't arrived, however, were over 200 newly built six-mule wagons that were somewhere west of Fort Abercrombie.[2] Each of these, and most of the wagons at Fort Rice, could haul 5,200 pounds of supplies. Moreover, further supplies were being sent upriver to meet the expedition where Glendive Creek in Montana empties into the Yellowstone River.

This chapter begins with the first of Barrows's letters to the *New York Tribune*, submitted June 13. Following Barrows is the answer to a question that nontechnical readers might easily have: what did a nineteenth-century surveyor do? Coincidentally, in August, the *New York Times* ran an article describing the surveying process. Milnor Roberts's letters to Northern president Cass relate to final preparations, and Edward Konopicky complains about drinking from the Missouri River.

Accounts differ as to the size of the surveying expedition's military component. Because of this, the official numbers submitted by Col. David Stanley should be considered authoritative.

---

[1] The 3" rifled "Rodman" gun had become a standard Union artillery piece by the end of the Civil War. Technically, however, the gun was not cast in the patented Rodman process yet somehow acquired the name.
[2] Fort Abercrombie was on the Red River in Dakota, thirty-some miles south of Fargo.

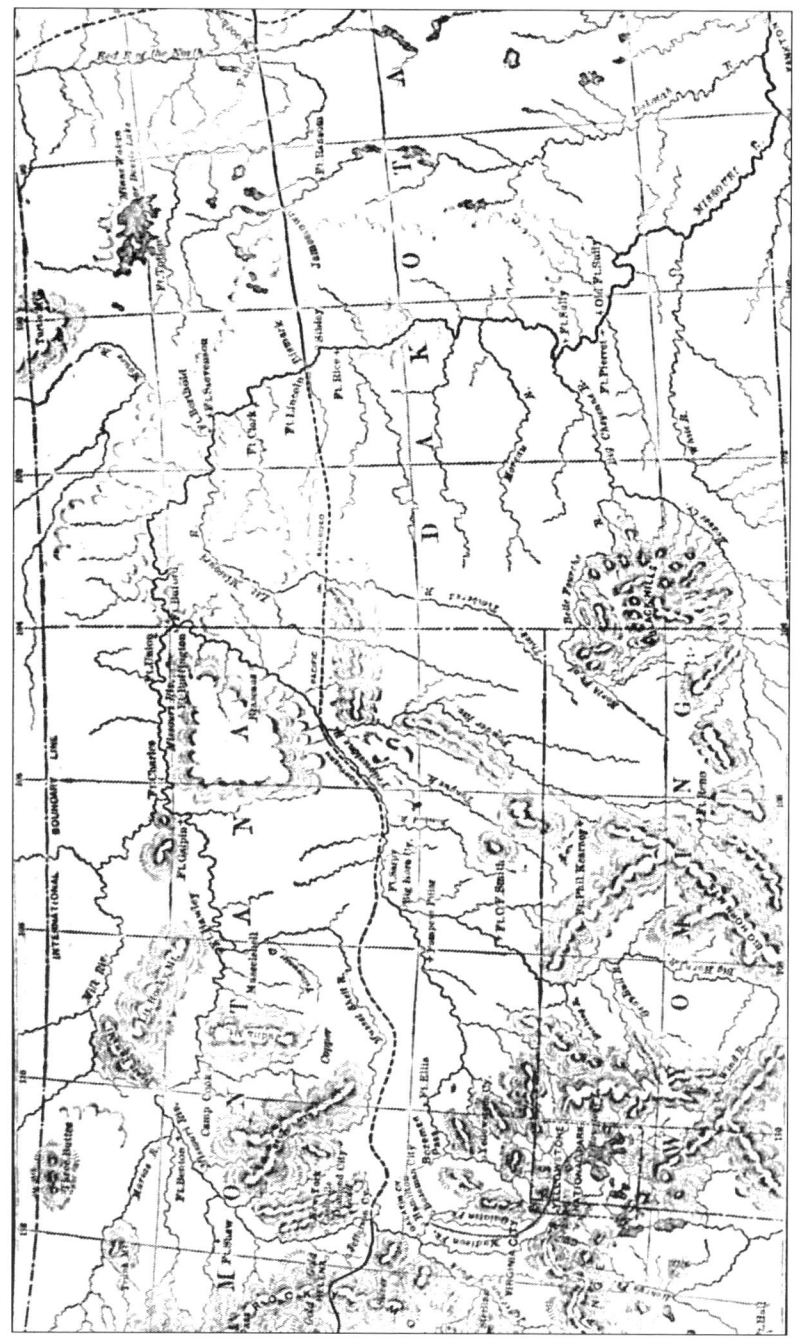

This six (of eight) column map was given first-page placement by the *New York Daily Tribune*, June 28, 1873, to assist readers with finding locations in Barrows's dispatches.

Samuel J. Barrows to the New York Tribune

THE GREAT WEST

LARGE MILITARY FORCE TO INTIMIDATE
THE INDIANS—DETAILS OF THE ORGANIZATION—
PROSPECT OF A RICH HARVEST OF
SCIENTIFIC TROPHIES.

FORT RICE, DAKOTA TERRITORY, JUNE 13.—This place is so remote from the East, and until recently so far removed from direct railroad communication, that it may be necessary to tell your readers how to get here. The most direct route is from New York to Chicago, from Chicago to St. Paul, St. Paul to Duluth or North[ern] Pacific Junction 24 miles west of it, and thence by Northern and Pacific Road to Bismarck, on the Missouri River. From Bismarck to Fort Rice it is 25 miles by land, and between 50 and 60 miles by river, owing to its great disposition to kink. Eventually there will be a much more direct route from St. Paul by the way of the St. Paul to Fort Breckinridge, and intersect the Northern Pacific Road at the Red River.

Within the last 10 or 15 years so many new Territories have been manufactured in the West, and so many old ones turned into States, that one has to be an active student of geography to keep up with the changes on our map. I doubt if many well read people in the East, even professional men, other than teachers, could give the boundaries of our Territories correctly without reference to a map. For the benefit of those who have neither the inclination nor the opportunity to consult the latest maps of the United States, let me say . . . that Dakota is bounded on the north by the British Possessions, on the east by Minnesota and Iowa, on the South by Nebraska, and on the west by Wyoming and Montana. The Red River Valley runs through the eastern part, and the Missouri runs diagonally across the Territory. Fort Rice is very nearly in the center of the Territory. Military forts have been established at convenient points on the river and through the eastern half of the Territory. Yankton, the capital, is in the south-eastern part of the Territory, on the Missouri. It has a population of about 2,200. Vermillion, in Union County, has about 1,000; Elk Point, Sioux Falls, and Springfield have about 500 each; Bismarck, 1,000. There are a score of other little places tucked down in the south-eastern corner, near the Iowa line. In all, the whole Territory has a population of about 25,000 to 30,000, confined to an area quite small when compared with the extent of the Territory. The far greater part of the Territory is a new and unsettled country. That portion of it west of the Missouri has scarcely been explored at all. The Sioux and other tribes of Indians at present occupy the larger portion of the Territory. Until they are subdued settlement in the interior cannot advance.

A NEW TOWN

Of the towns mentioned above, Bismarck is the only one with which I have a personal acquaintance, and that has been limited to one night.[3] Could the great German after whom it is named learn of the honor that has been conferred upon him he would be sadly lacking in the sense of humor if he did not enjoy a good laugh. This is the only satisfaction I think he could derive from the compliment. But as he is never likely to hear of the place, I fear that this satisfaction will be denied him. But even Bismarck the great was once little, and perhaps the time will come when Bismarck the little will become great. The town thus named is situated on the Missouri River at the Western terminus of the Northern Pacific Road. The place dates its existence from July, 1872, and is thus hardly a year old. It claims about a thousand inhabitants, and I should suppose now two or three hotels; several stores, and that essential feature of a Western town—indefinite number of rum shops. It has not a very good reputation in respect to morality, but I think it is not lacking in hospitality. I arrived there late at night. One half of the town had gone to bed, and the other half were in the rum shops. The hotel was full, and closed for the night. There seemed to be no resort but to return to the car we had left, and pass the night, but a fellow-passenger succeeded in arousing a store-keeper known to him, and we were kindly taken in and quartered in the store, and left on a bed of buffaloes and blankets. Whatever may be said against the Western character the charge of inhospitality cannot be brought to its households. The town is now rejoicing over the completion of the road clear to its doors. The road was only brought to this point on Thursday last, a week ago today. The town was built on faith of this termination. Some dissatisfaction is expressed at the way the Northern Pacific road handles the town sites. Men are afraid to buy, they say, till they know whether the town is to remain here or not. The nearness of Fort Rice to Bismarck, and the fitting-out of the

---

[3]The Northern Pacific, in its efforts to raise German money, officially named its city after Otto von Bismarck, chancellor of the German Reich in April 1873.

Yellowstone Expedition, has accelerated the trade pulse of this infant town considerably. The store in which I slept was opened a few days since. The first day the owner sold $584 worth of goods. The man would be a Stewart in a few years at this rate of increase.[4] I imagine, however, that when the Yellowstone Expedition leaves, things will be rather dull in this vicinity. At present, it is the all-absorbing topic through Dakota and Minnesota, in social, military, and commercial circles.

A SUPPLY OF PORK—PROSPECT OF INDIAN FIGHTS.

The boat which I was able to take from Bismarck was detained several hours to await the arrival of Gen. Terry, commander of the Department of Dakota, who in company with his staff, has come down to confer with Gen. Stanley, and witness the departure of the expedition. Our boat was one of the type of Mississippi steamboats, which have figured so much in the history of internal navigation in this country. . . . The boat, I have reason to believe, was prepared for any . . . emergency. Pork fat is historically an essential element in any steamboat contest on the Mississippi or Missouri. There seemed to be a plentiful supply of this material on board. Being so far up the Missouri, having the river mostly to ourselves, the engine fires had little occasion for its use. The superfluity was therefore daily served up three times on the table. Fried, boiled, smoked, and melted down into liquid fat, the extra hog was the regular digestive staple. It was fried hog for breakfast, boiled hog for dinner, and smoked hog for supper. A stream of melted hog infected the gravybowl. The vegetables, the hominy, and the corn were saturated with the same material. After using this fuel for five successive meals, I heartily wished that some rival steamer might overtake us and stimulate our engine fires.

Fort Abraham Lincoln you will probably not find on the maps. It is about three miles down the river from Bismarck on the Western side. It is situated on a high bluff and commands an extensive and beautiful view of the surrounding country. There are at present three companies of infantry there, under command of Lieut.-Col. Carlin. We stopped three or four hours at the fort to give the officers on board a chance to fraternize with the officers of the post. Many of them had not seen each other for years, and four hours was little enough to renew friendships, recount experience, and fight over old battles. I learned something of the nature of the Indian operations along this side of the river. The Sioux

---

[4] Alexander T. Stewart (1803–76) was the "father" of the department store business.

are proving very troublesome. For a month or two past they have been growing more and more audacious. It is the opinion of officers here that nothing but a good thrashing will check them. The recent statement in the papers, however, that the Indians had attacked the Fort in numbers is not correct. About 40 or 50 Ree Indians are employed at the fort as scouts. They are brave and effective men, but are, with their whole tribe, deadly enemies of the Sioux. The Ree scouts take the cattle out on the hills near the fort. The Sioux, who are well armed and mounted, maintain a constant watch to attack them at a disadvantage. If they see a few Rees a little removed from the Fort, they will frequently charge upon them and drive them in, usually being careful to keep out of musket range of the Fort. Sometimes they appear in bands of 50 to 100 men. The long roll is then sounded, and the garrison called out. But as the garrison is composed entirely of infantry, the Sioux are quite indifferent to the array. They know the soldiers cannot pursue them, and so they leave at their leisure. It was a raid of this kind that occasioned the report of the attack on the Fort. If the Ree scouts were removed the Sioux would probably not trouble the Fort. A company of cavalry then could scatter them very effectually. After the expedition returns Gen. Custer's cavalry will probably be stationed at Fort Lincoln, or in its vicinity. In one of these recent Indian raids the artillery at the Fort did good service. A well-directed shot killed two of the Indians and hastened the departure of the rest of the band. Since that time they are more careful how they come within range.

INTENTIONS OF THE EXPEDITION.

Leaving Fort Lincoln at midnight, we reached Fort Rice in the morning about six o'clock. Gen. Terry was received with the usual military salute and conveyed to headquarters. To-day the various battalions will be inspected in succession by Major Lewis of Gen. Terry's staff. A conference will be held with Gen. Stanley in relation to the expedition.

By the time this reaches you the Yellowstone Expedition, concerning which so much has been said and so much left unsaid, will have departed. According to programme the departure ought to take place by the 15th inst. But . . . [s]upplies have failed to arrive at the time expected. The wagon train is yet 80 miles from here, and cannot arrive before Sunday night. As the expedition is going so far from its base of supplies a few days' time can better be spared than many days' rations. An expedition so large as this involves a great amount of preparation. If it is able to leave by the 25th it will do well.

In view of the early departure of this important expedition a full account of its objects, organization and route will form a fitting introduction to a faithful chronicle of its progress and results. It is known here and all through the West as the Yellowstone Expedition, because one of its objective points is the Yellowstone River, and much of its operations will be conducted along its valley. But it is identical with what in the East has been called the Northern Pacific Railroad Expedition, because under its protection the engineers of that road will continue their survey.

The Expedition has several related objects. First, as already intimated, to serve as an escort for the survey and location of the Northern Pacific Railroad, which will begin at the Missouri River and be extended to the interior of Montana, to connect with the survey from the Western coast. Second, to intimidate and subdue the Indians, who are bold and hostile, and very much opposed to the introduction of this road. If the Indians are quiet and inoffensive, their peace and tranquility will not be disturbed. Should they be disposed to be hostile, the expedition is fully prepared to meet such a disposition. It will be a part of the military plan to make a thorough determination of the whole section of country through which the army will march, with a view subsequently to establish two new military posts in the North-West.

It is probable that one of these will be located on the Yellowstone, at the mouth of the Powder River.[5] An appropriation of $200,000 was made by Congress to establish these two posts on the line of the Northern Pacific Railroad. The appropriation was made on the recommendation of Gen. Sheridan. A third and important object of the expedition is to make a thorough scientific investigation of the traversed region, which has never yet been explored, and which it is expected will furnish a rich field for study and research, especially in its zoological and geological features. The large wagon train which takes out forage and supplies will return almost empty. This furnishes an excellent opportunity for collecting and forwarding to the East the large amount of specimens which it is hoped the region will afford.

Previous expeditions of this sort, while they have engaged the services of eminent scientific men, and have been valuable in results, have been defective in transportation. It is especially distressing to a scientific man

---

[5] The posts were never built.

to find extensive and valuable material on every side, and yet have no means of bringing it away. The best organized scientific expeditions are not able always, in the field, properly to classify and differentiate the multitude of specimens that may require observation and reference. Generally on returning, collections are divided and submitted to the inspection and criticism of eminent scientific specialists, who are thoroughly versed in their respective departments. Whatever may be the character of the field-work done on this expedition—and there is no occasion to suppose that it will be done otherwise than well—if extensive collections are made in the various scientific departments and returned successfully to the East, the expedition cannot fail to be fruitful in scientific results.

THE MILITARY ORGANIZATION

Gen. Stanley's staff is as follows: H. H. Ketchum, Adjutant 22d Infantry, Acting Assistant Adjutant-General; P. H. Ray, Second Lieutenant 8th Infantry, Chief Commissary of the Expedition; Capt. Edward Baker, Chief Quartermaster; Assistant Surgeon J. P. Kimball, Chief Medical Officer; Lieut. Jas. H. Jones, Acting Aid-de-Camp.

The expedition will leave Fort Rice supplied with 60 days' subsistence and forage. Three steamers will take supplies up the Missouri and Yellowstone to the mouth of Powder River, where the expedition will meet them. One of these steamers will probably be used to ferry the command and supplies. It is possible that on the return of the expedition the infantry may be conveyed to Fort Rice on boats. The troops and officers are limited to the smallest possible quantity of personal baggage. The Department prescribes for the use of enlisted men one overcoat, one blanket, two shirts, two pair drawers, five pair socks, two pair shoes, and one poncho (rubber garment). In addition one pair of shoes per man will be carried in bulk on the wagons; the other clothing in the knapsacks. For the cavalry part of the command a corresponding allowance will be taken in the saddle bag. Each officer will take a valise; trunks and other *impedimenta* will be judiciously eschewed. The ammunition allowance is fixed at 200 rounds per man, cavalry and infantry, 40 rounds to be carried habitually in the cartridge boxes, the remainder in bulk in the wagons.

Thirty Indian scouts and eight half-breeds mustered as guides, and scouts will accompany the expedition. The 30 Indian scouts are enlisted men, and receive regular soldiers' pay. The half-breeds are employed at the rate of $75 a month. Reynolds, a well-known hunter, who is one of the most skillful men in his calling in the North-West, will accompany the expedition with some assistants, and add as much game as possible to the stores.[6]

Gen. Stanley, who commands the expedition, is one of the most efficient officers in the service. It is only necessary to refer to his army record. He graduated from West Point in 1852, and has been in the service ever since. During the war he was a Major-General of volunteers, and commanded the Fourth Corps operating in the West. He is closely identified with several of the most important military successes in that department. At the close of the war he was brevetted Major-General of the regular army. His present command, independent of the expedition, is what is called the middle district of the department of Dakota, which comprises . . . Fort Buford to Yankton, and includes Forts Buford, A. Lincoln, Rice, and Sully—in fact, all the military posts on the Missouri above Yankton.

Gen. Stanley's devotion to the army may be seen from the fact that in his whole term of service of 21 years, he has had but 30 days' leave of absence, and that was on the occasion of his marriage. Army officers are allowed to have 30 days' leave of absence in each year, and are generally not loath to take it. During the war he was wounded two or three times, had his leg broken and his horse shot under him. Gen. Stanley is between 45 and 50 years of age.[7] His 20 years of active service have bronzed his features and silvered his hair; but he is good for many years of work yet. Last year he commanded the escort to a similar party of surveyors of the Northern Pacific Railroad. It was composed of 12 companies of infantry, without cavalry—in all, between 700 and 800 men.[8] They were gone a little over three months. The expedition last year went as far as the mouth of the Powder River.

ORGANIZATION OF THE SURVEY.

To give an idea of the work of the present survey let me say what has already been done on the Northern Pacific road. The construction of the Northern Pacific Railroad was begun in 1870, and was completed

---

[6]"Lonesome" Charley Reynolds (1842–76) had been on the 1872 eastern survey. He died with Custer three years later.

[7]Stanley, forty-five, was born June 1, 1828.

[8]The real number was closer to 575 officers and men, with teamsters, surveyors, scouts, and so forth, bringing the total to just over 700.

across the State of Minnesota in 1871—a distance of 228 miles from the junction with the Lake Superior and Mississippi Road, or 252 miles from lake navigation at Duluth. The section west of Minnesota between Red River and the Missouri was put under contract in October, 1871. As already noticed, the last rail on this section of the road was laid last week, on the 5th inst. In 1871, a preliminary survey was run clear across the continent. The survey, in 1871, commenced at the mouth of the Heart River, opposite Bismarck, on the Missouri, and proceeded 150 miles west. From there it descended to the Little Missouri River, following a stream which was named Andrews Creek. Following down the Little Missouri five miles, they took another stream called Davis Creek,[9] and followed it about 21 miles across the divide between the waters of the Little Missouri and Glendive Creek. They followed the Glendive to the Yellowstone, on the 47th parallel, which ended the survey.

Last year they began the survey about seventy-five miles west of the Missouri, and bore to the south-west, leaving the line of the previous year a distance in the aggregate of forty miles north, and made their objective point the Yellowstone, at the mouth of the Powder River.... This route was rejected, however, by the Board of Directors as impracticable.

This year it is expected that a decided location of the line will be made from the Missouri River to the mouth of Glendive Creek. The survey will be continued up the Yellowstone. A large portion of the country from Washington Territory east has been surveyed. The eastern and western surveys are about 150 miles apart. It is the programme now to follow up the Yellowstone and Pryor's Rivers and cross the divide between the Yellowstone and Missouri Rivers to the Musselshell River.[10] Following that on a line due east, they will connect the two surveys from Puget Sound to Lake Superior.

THE SCIENTIFIC PARTY.

The scientific party is composed as follows: J. A. Allen of the Cambridge [Harvard] Museum, in charge of the zoology, botany, and palæontology, and chief of the party; Dr. L. R. Nettre, mineralogist and geologist; W. R. Pywell of Washington, photographer; Edward Konopicky of

---

[9]Davis is on the eastern side, Andrews the western, with Barrows confusing the two.
[10]The actual plan was to cut north to the Musselshell at Pompey's Pillar, some fifteen miles below Pryor's Creek, not Pryor's River.

Cambridge Museum, artist; C. W. Bennett of Cambridge, taxidermist and general assistant.

The scientists came down from Fort Lincoln with a wagon train, on Sunday last, with an escort of forty men and four Indian scouts. They are now comfortably quartered here, have a pleasant encampment, and all their outfit secured. Major H. M. Lazelle of the 8th Infantry, an officer of scientific tastes and a thorough scholar, has been detailed to look after the wants of the scientists. He is himself the author of a recent work entitled "One Law in Nature," which is exciting a good deal of attention in scientific and critical circles.[11]

The following order will explain just what arrangement has been made for the scientific party. It will be seen that the provision is generous and adequate:

WAR DEPARTMENT, WASHINGTON, D.C., MAY 6, 1873.
Col. D. S. Stanley, Fort Sully, D.T. through Lieut.-Gen Sheridan.

COLONEL: The expedition under your command, organized to escort the survey of the North Pacific Railroad, and to protect the location and construction parties of that railroad west of the Missouri River, will make a thorough exploration of the region traversed, collecting and reporting all obtainable information in relation to the natural resources and capabilities of the county. For this purpose all officers and persons on the expedition are charged with the duty of contributing as much as may be in their power to aid in the collection and preservation of this knowledge.

The commander of the expedition will be in charge of this as of all other departments of his command. He will have, for the better execution of this duty, a corps of scientists specially skilled in the several branches of knowledge, who will be under his special direction and protection, who will accompany the expedition and will be provided with rations, transportation, camp equipage and necessary assistance by the officers of the appropriate departments of the army. They will be appointed by the Secretary of War, and will receive special instructions in their several spheres, prepared under the advice of the National Academy of Science. These gentlemen, upon the return of the expedition to the Missouri, will be furnished by the Quartermaster's Department with transportation to their homes, where each will make up his personal report and forward it to the Secretary of War. The collections of the expedition will be transported by the Quartermaster's Department to Washington, and delivered to the National [Smithsonian] Museum for classification, arrangement, and preservation, as by law required.

A list of the names of the gentlemen invited to join the expedition is

---

[11]Barrows's comments were generous. *One Law in Nature* was roundly panned at the time.

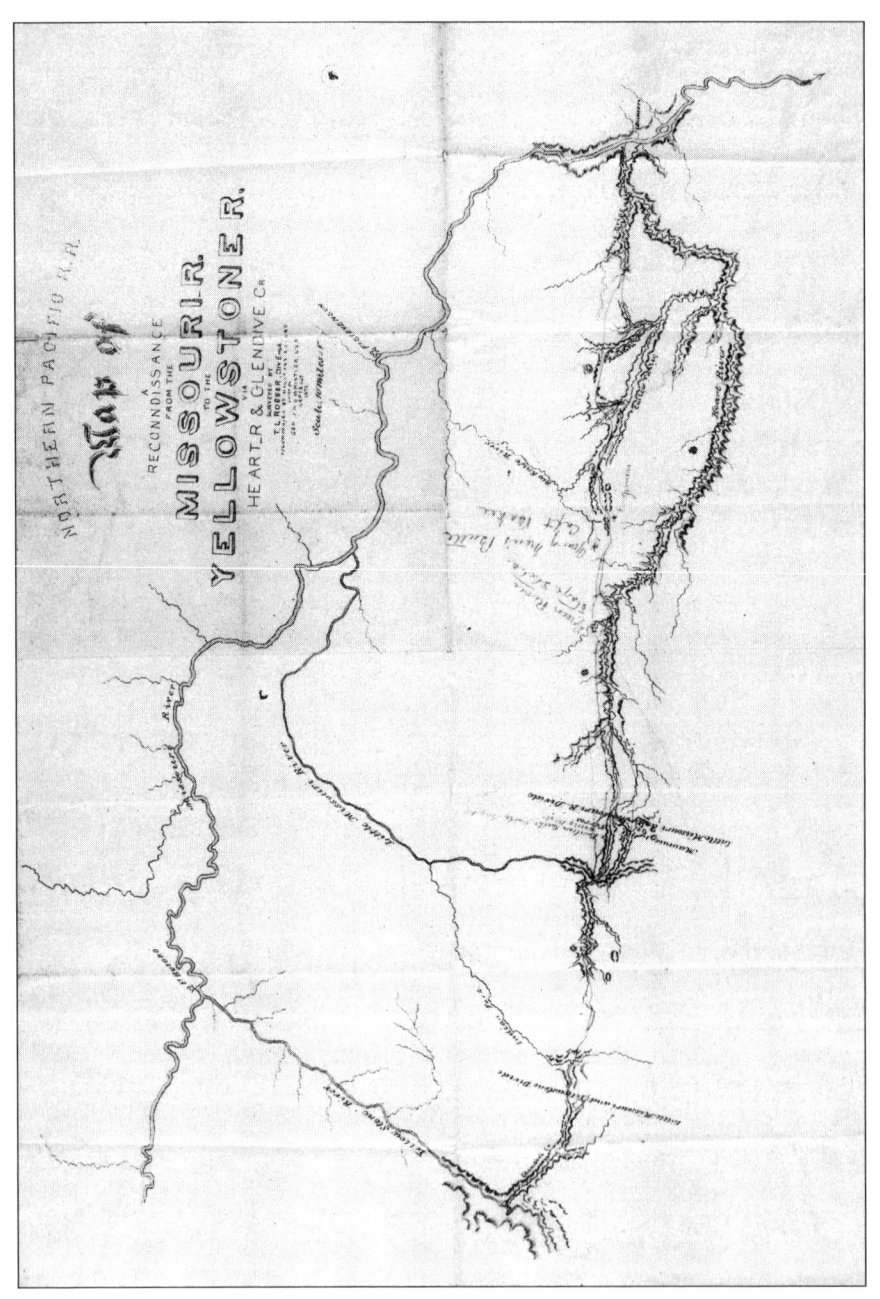

enclosed, designating their special spheres of inquiry. If any of them should be unable to join, others will be invited in their places, and of all such changes you will be duly informed. A copy of their instructions will be forwarded to you.

These gentlemen and their objects are commended to your care and protection, and you are instructed to direct all officers and men under your command to aid them in their special duties, and to contribute to the success and completeness of the exploration, as far as may be possible, without injury to the special duties of each. You will detail a sufficient number of troops, commanded by a young, active, and intelligent officer, and who takes an interest in scientific pursuits, to attend specially to the members of their corps....

[Signed by the Acting Secretary of War]

Five hundred dollars has been allowed for tools and special outfit for collecting, and $1,000 for photographic apparatus and material. The photographer [Pywell] is supplied with a spring wagon for transporting his material and apparatus, and a spring wagon is also supplied for the other members of the corps. A saddle-horse is supplied to each member, and a heavy wagon and team will transport the baggage.

BEGINNING OF THE SURVEY.

The railroad surveyors begin their work this year at the mouth of the Heart River, which is nearly opposite Bismarck, and some 25 miles from here. In order that the delay in the departure of the expedition may not retard the survey, Gen. Stanley has ordered Major Townsend of the 9th Infantry, with four companies of the 25th Infantry, and a detachment of 25 men and a commissioned officer from the cavalry,[12] to proceed at

---

[12]Lt. Charles C. DeRudio (1832–1910). One of the Seventh Cavalry's most interesting officers, DeRudio had attempted to assassinate French emperor Napoleon III fifteen years earlier. He was spared the guillotine at the last minute through the actions of his beautiful wife with the emperor (one might guess how). He then survived the Little Big Horn (by hiding) and lived to a ripe old age.

(opposite) TOM ROSSER'S BISMARCK TO GLENDIVE CREEK MAP,
USED IN 1871 AND 1873

Made of a water-resistant, parchmentlike material, Rosser's map is more than five feet wide. The Heart River (dominant markings in lower half of map) was the surveyors' 1871 route, while in 1873 Rosser took a basically straight line that eliminated more than twenty miles of track by skirting the Heart River's southern bulge. *Courtesy Albert Small Special Collections Library, University of Virginia, Charlottesville.*

once to Fort Lincoln, then cross the river and proceed with the surveyors along the Heart River. They start tomorrow, the 14th inst., supplied with 15 days' rations and 100 rounds of ammunition per man. The main body of the expedition will move almost directly west from here, and overtake the detachment in two or three days on the Heart River.

Published June 28, 1873

New York Times
HOW RAILROADS ARE BUILT

AUGUST 12, 1873.—[Once agreeing upon a general route] . . . the next thing is to ascertain over what line of country the road should be located, and what it will cost to build. The consideration which mainly determines the location of the road is grade; the steepness of its ascents or descents. As railroads are at present operated, it is found highly desirable not to exceed a certain steepness of grade—say fifty feet to the mile. Beyond this the cost of operating the road is greatly increased, and its carrying capacity greatly diminished.

As a general thing, it will be found that a country intervening between any two distant points is considerably diversified and broken up by hills, valleys, rivers, etc., so that it is impossible to run a straight line without encountering obstacles or submitting to enormous "cuttings" and "fillings," and building tunnels and bridges. All these involve great expense, and the engineer avoids them by judicious location.

A chief engineer being selected [e.g., Rosser], it becomes his duty to ascertain the best location of the road, yet regard has to be for the greatest economy. The road must thread the country, crossing the hills at the lowest points, and the streams where narrowest. Rocks, involving heavy cuts or tunnels, must be avoided, as also places where the road would be of difficult maintenance. At the same time the line must not be taken too much out of the way or an increase in distance may more than counterbalance the saving. It would seem nothing short of inspiration or instinct could lead him directly to the best line.

However, a practical knowledge of physical geography guides us to those localities where we encounter the fewest obstacles. The water courses of a country always indicate the line of most uniform grade, and water always seeks the lowest points. Streams are the locating engineer's best guide. If the points are on the same bank of a river, there is generally little doubt as to the proper location of the road. So much are water courses a guide that an engineer can nearly always pick out the general line of his road on a map. If on a map he sees that his line is to run in the same direction as the principal water courses, he will expect to find light work.

If, however, he sees that [his line] runs across the direction of the main streams, he looks for great trouble in the location of his line and heavy work, for he must traverse the intervening ridges, which form the natural divisions between the streams. These indications are only general, and the ground over which the road is to pass has to be minutely examined.

In the first place, the chief engineer makes a preliminary survey or reconnaissance of the ground. He provides himself with the best map possible, and avails himself of what local information he can obtain as to gaps in the hills, low points in dividing ridges, etc. This is a part of the profession which cannot be learned from books. It depends on experience, and a kind of heaven-born instinct. In fact, as our engineer represents a type of peculiarly gifted being [let us] sketch his portrait.

He is a man endowed with intuitive faculties, who feels the right thing to do.[13] He is possessed of a certain professional conscience which whispers in his ear what to do, without always telling why. He is quick, resolute, determined and possesses great confidence, and is, generally speaking, rather conceited. Candor compels us to add that he almost invariably drinks, not necessarily to excess but tolerably hard.[14] He also generally chews and smokes.

After this statement one word to the younger members of his corps. The weaknesses of men in high positions are more easily imitated than their great qualities. Let the young engineer, rest assured that were he to drink himself drunk every night, and smoke and chew himself to a skeleton, he would not thereby insure his early promotion.

This preliminary survey made, and the location of the line selected, the next step is for an engineer corps to "try the line" by critical instrumental tests. However skillful the chief engineer may be, he can never tell without such tests whether the ground is really what he supposes.

---

[13]Rosser's intuition to follow Sweet Briar Creek rather than the Heart River's Great Bend is a perfect example.

[14]Could a description of Rosser be more on the mark?

The object is to get the profile and alignment of the route. The profile is a vertical section of the line, showing the ups and downs, or differences of level of its various points; and the alignment is a plan or map showing the deviations of the route from a straight line and its length.

An engineer corps, for location, consists of an assistant engineer, a transit man, leveler, rodsman, two chainmen, one or two flagmen, and a gang of axmen. If operating in a sparsely settled country there are a cook, teamster, and generally one or two boys, who, in the intervals of skylarking, are supposed to make themselves useful.

Work is conducted this way: the principal assistant has charge of the whole party and, in a difficult country, keeps well ahead to select proper ground for the line, communicating constantly with his transit man.[15] The line being started in favorable ground, is continued in a straight line as long as possible. Stakes are driven every 100 feet and numbered consecutively from 1 upward.[16] The transit man, looking through the telescope of his instrument, takes care that these stakes are driven in a straight line. When it becomes impossible to continue further, a deviation is made to reach better ground, the transit man noting the station at which the deviation is made, and the degrees and minutes of the "deflection angle" with its original direction. These data enable draughtsman to prepare the alignment or plan of the route.

In a preliminary survey, the "levels" are the most important points. These are taken by the leveler and rodsman. These two follow the transit man, and ascertain the difference of level of each station, as the stakes mentioned above are called. Then the difference of level is taken. From this data, the profile is plotted. It is evident that if the profile were drawn on a large scale, small differences of level would be indiscernible, while a scale sufficiently large to render these differences perceptible would give an enormous length. The difficulty is got over by use of "profile paper," by which the profile is plotted on two different scales, one for the horizontal measurement, and one for vertical distances.

This profile gives the engineer an exact representation of the obstacles he must encounter. It shows exactly what cuts and fills he will have to

---

[15] Transitmen used theodolites, an instrument with a telescope mounted on a three-legged stand. Theodolites measured horizontal and sometimes vertical angles. Invented in the 1500s, the type that Northern Pacific surveyors used dated back to the 1780s.

[16] This is incorrect or at least concerning the Northern Pacific's surveyors. Measurements were made using a sixty-six foot Gunter's Chain, invented in the early 1600s and the basic U.S. government measurement instrument dating back to 1787.

make in running that line. It is very seldom that the preliminary line is found to answer the requirements of a permanent location. [Thus] it is often necessary to run several separate trial lines before one is found. Fifty miles of trial line to secure ten of location is by no means uncommon.

In running these trial lines rough work is encountered. You must go where the line takes you; through swamps, tangled underbrush, [and] deep streams. Hard work at that, for there is always a rush to get the trial lines run as quickly as possible. But oh! great is the peace when the word is given to knock off for the day, and the blazing fire becomes visible surrounded by savory messes. No sumptuous banquet at Delmonico's, no turtle soup, no Sillery frappé, no anything but fried ham, slacked-baked bread, boiled rice, and black coffee....[17]

Roberts to Cass

Bismarck
June 14, 1873

I left here on Wednesday afternoon 11th inst. in company with General Terry, and others, and went with them on board the steamer *Miner* which had come up. The *Miner* belongs to Dr. Burleigh[18] and had been chartered at $500 per day to keep along with the 7th Cavalry which arrived on the 10th at Fort Rice. Mr. Mead brought the General and party out early, leaving Fargo at 5 A.M. and arriving at Bismarck about 5½ [P.M.]....

I paid my respects to General Terry, General Crittenden, who commands the Fort, General Stanley who commands the expedition, and Genl. Custer who commands the 7th Cavalry....

Lt. Col. Grant also called today, just before starting for Fort Rice. He looks much stouter and heavier than when I saw him passing his examination at West Point two years ago.

Source: Northern Pacific Records, Secretary's Unregistered Letters, at the Minnesota Historical Society, St. Paul.

---

[17]Sillery frappé was a popular French chocolate drink made with champagne. "Slack-baked" means half-baked.

[18]Walter Burleigh (1820–96) was active in Territorial politics. The year 1872 saw his company receive the Fargo-Bismarck construction contract. In turn, he sold it to another company, a typical Crédit Mobilier–type kickback. As a Missouri River steamer cost some $25,000, $500 a day was high.

[This letter continues the next day:]

I succeeded in getting so much [done] yesterday, but did not close my letter as no mail would leave here till tomorrow morning. In the afternoon the mosquitoes became so bad that it was next to impossible to write, and it is not much better today. I had made an arrangement . . . [with] Capt. [Grant] Marsh[19] [who] agreed to lend us a pilot engineer and with an improvised crew we had the boiler of the *Ida Stockdale* cleaned, pumped up, and some wood put on board, and got steam up. By this time it was dark, and we postponed our ferrying till this morning, when we put all on board and shipped them out to Fort Lincoln, whence the party is to start tomorrow on the survey.

General Terry and General Stanley, finding the main body of the wagons were behind time, agreed to send four companies of infantry up from Fort Rice to Fort A. Lincoln so as to be here on time to start on the 16th. . . . These are to start out with the engineering party, while the main body, including all the cavalry, will go from Fort Rice and intercept the engineering party some days out. . . .

The Government Officers, all of them, are anxious to do everything in their power to facilitate our operations and I am disposed to do all in my power to aid them.

Bismarck
June 16, 1873

The following is a list of the Engineering party engaged in the survey west of the Missouri River which began field work today.

General Thomas L. Rosser	Engineer in charge	$300[20]
A. O. Eckelson	Chief of Party	$175
M. Meigs	Transit & 1st Ass'nt	$150
H. M. Reed	Assistant Transit	$125
H. E. Stevens	Leveler	(not stated)
P. T. Winston[21]	Leveler	(not stated)

---

[19]Grant Marsh (ca. 1834–1916) was one of the upper Missouri's leading steamboat captains and had a long association with the army.

[20]Some salaries are drawn from a letter that Rosser sent to 1870–72 surveyor Edward C. Jordan (May 21, 1873) inviting him for another summer. Author's collection.

[21]Fendall G. Winston (1849–1928) and Thomas W. Winston (dates unknown) were younger brothers of Phillip B. Winston and "Lizzy" Rosser.

A. O. Dahna	Topographer	$125
D. Loring	Assistant Topographer	(not stated)
A. S. Berry	Rodman	$75
F. G. Winston	Rodman	$75
Zack Taylor	Rodman	$75
A. Fry	Flagman	(not stated)
A. Swanson	Chainman	$60
Iver Wicke	Chainman	$60
E. P. Deerfield	Flagman	$60
P. W. Lewis	Stakeman	$60
T. M. Winston	Axeman	$60
I. Bates	Axeman	$60
H. Neilson	Campman	$80
P. Mulloy	Cook	$60
J. McGuire	Assistant Cook	$50
Loomis Babb	Wagon Master	$80
J. Barr	Teamster	$50
P. Rapp	Teamster	$50
T. Reid	Teamster	$50
W. Hammond	Teamster	$50
R. C. Watson	Teamster	$50
A. J. McDonald	Teamster	$50

Some of our appointees did not make their appearance, but we were able to fill up the party with good men. It is a first rate working party. In the hurry of getting off I did not get the exact rates of pay but . . . [it is] about $ 2,210 exclusive of General Rosser's. . . . He is worth more and I think could command it elsewhere. . . .

They will have several newspaper reporters with the expedition, I believe; and, almost at the last moment, came Professor [William] Phelps from Vienna,[22] under the auspices of Senator Windom; presented in such a way to General Rosser, that I believe he *goes*—furnishing his own horse and saddle; being sheltered and fed, himself. As to engineering matters, he is as green as a girl . . . but he may be of real service.

The newspaper reporters are to me regular pests. They are either ignorant or impudent, or the two combined.

---

[22] Roberts confused Phelps with Konopicky.

[P.S.] Tuesday night. I was about closing my letter when word came of a fight with the Sioux over the river, and as several newspaper reporters were around, I jumped on a horse [and] rode down to the ferry and walked up to the Fort in double quick time. I found Genl. Carlin and got from him the particulars for my long telegram of this date. When I returned, sure enough, telegrams had started which Mr. Mead stopped at Fargo and ordered that I should see them before they were sent. Afterwards I looked over them and as they tallied with the *facts*, I said so to Mr. Mead. We need not be afraid of plain facts. There will be parties enough to exaggerate and misrepresent. There is no particular fun in such close fights.

St. Paul
June 22, 1873

I had the pleasure to escort safely from Bismarck to St. Paul, General Terry and staff, and six ladies, wives of officers of the 7th Cavalry, whose husbands have gone with the Yellowstone Expedition.

General Terry spoke to me concerning our crossing of the Missouri River. I then ascertained that his sole object was in relation to the location of the cavalry post at Fort Abraham Lincoln. . . .

Edward Konopicky to His Parents

Fort Rice, Dakota Territory
June 17, 1873

Today is already the 8th day that I have been enjoying camp life. . . . Life in the outdoors suits me well, as it does the others. We all look healthy and so tanned that we are beginning to look almost like Indians. Not surprisingly, the heat here is almost unbearable, and the hot months, July and August, are still to come. On top of that, our route to the Yellowstone leads through, in parts, a completely treeless, rocky, and barren country. We will start at the end of this week, and from here to our destination, we have to put about 50 English miles behind us.

For three days a big Missouri steamboat has been busy ferrying 200 wagons, each drawn by 6 mules, across from the opposite shore. Among the wagons were the spring wagons meant for us. We expect our horses any hour. The huge column, including infantry and cavalry, will probably stretch 3 to 4 English miles. The Indians will be sorry. (The Indians along the Yellowstone River are almost all unfriendly tribes who have refused to make peace with the Whites.)

The five of us [the Scientific Corps] are well taken care of, due to the fact we were sent by Washington, and enjoy the highest attention of the staff officers who visit us daily. The Commanding General is Stanley. Our food is as good as is possible in a camp, and includes fresh and smoked pork (which is not a special joy 3 times a day), tongue, vegetables including corn, cabbage, tomatoes, French beans, pickled cauliflower, various preparations from cornmeal, fresh bread, zwieback hardtack, butter, preserves of strawberries, peaches, coffee, and tea. We bought the provisions. The worst part of camp life is the water. The only water we have is from the Missouri and it is like the seepings from a brick oven, except a little thicker. The water has to stand for several hours before most of the sand settles. It is also warm. You can well imagine with what longing I think of beer and wine.

I haven't seen any since St. Paul and I must do without for another 4–5 months. Therefore, just think, I have taken up smoking. I smoke one or two cigars each day. I laid in a supply of 100 cigars in Boston. I understand we will return via the Upper Missouri, which I like very much. . . .

I have begun my work with sketches of Fort Lincoln, Fort Rice, and a few Indian graves. A correspondent for the *New York Tribune* [Barrows], who is also with the expedition, requested sketches of the trip to the Yellowstone and there is a possibility of a book. I don't know if I will have time for him. After all, my job is to draw for Washington. . . .

Col. David S. Stanley's Memoirs

. . . The composition of the [Yellowstone Expedition] was as follows:

Ten companies of Seventh Cavalry, Lieutenant Colonel G. A. Custer commanding.

Ten companies of Eighth and Ninth Infantry, Lieutenant Colonel L. P. Bradley, Ninth Infantry, commanding.

Three companies Seventeenth Infantry and one company Sixth Infantry, Major R. A. E. Crofton, Seventeenth Infantry commanding.

Five companies Twenty-second Infantry, Captain C. J. Dickey commanding.

Detachment of twenty-seven Indian scouts, Second Lieutenant D. H. Brush, Seventeenth Infantry commanding.

One company Twenty-second Infantry (E) was organized as pioneers; two artillery squads, manning two 3-inch [rifled] Rodman guns, being selected from the same company.

As part of the force I was authorized to hire scouts, guides, and interpreters, and employed seven half-breeds in that capacity. The transportation, including every wheeled vehicle, amounted to 275 wagons [six being] ambulances.[23] The civilian employees numbered 353 men. The number of mules and horses to be foraged was 2,321.

The expedition was ready and started on the 20th of June; the effective force that morning being 79 officers, 1,451 men [and] a large herd, 700 head of beef cattle.

Four days previous to the main force leaving Fort Rice, a detachment of four companies of the Eighth Infantry, a squad of one officer and 25 men Seventh Cavalry, and one company. Sixth Infantry, from Fort Abraham Lincoln, had been directed to escort the engineering party from the crossing of the railroad on the Missouri River [at Bismarck]. . . .[24]

Source: Stanley, Personal Memoirs *(1917), 244–45, 247.*

---

[23]Not included were almost a dozen wagons used by the Northern Pacific surveyors, correspondents, and other individuals, including the two English noblemen.

[24]There were some 250 soldiers in this escort whom Stanley likely included in his totals.

CHAPTER 3

June 18 and 19

The arrival at Fort Rice of two hundred newly constructed six-mule wagons signaled that the Yellowstone Expedition would soon be leaving.[1] Nevertheless, the surveyors, unable to wait, took the *Ida Stockdale* to present-day Mandan, opposite Bismarck.

The surveyors were guarded by four companies of infantry (some two hundred men) and a contingent of twenty-five cavalrymen under 2nd Lt. Charles DeRudio. Within an hour of following the Heart River and before reaching Sweet Briar Creek, they were attacked in a woefully executed ambush by seventy-five to one hundred Hunkpapa Sioux. The colonel in charge of the troops, Maj. Edwin Townsend,[2] a man with an undistinguished combat record, was apparently flustered and relied on Rosser. "Engineers to the rear! Engineers to the rear!" Rosser bellowed, galloping to the front, taking de facto command and forgetting the promises he'd made to his wife about staying out of trouble.

Not until two companies of infantry from Fort Lincoln, Ree (Arikara) Indian scouts, and a few dozen belligerent (heavily armed and drink-fortified) civilians were ferried across the Missouri did the Hunkpapa retreat. They left behind two dead, both of whom were quickly butchered, then ritually eaten by the Rees.

In addition to Barrows's report, this chapter contains the diaries and letters from Luther Bradley and Rosser, who downplays his participation.

One document, an almost businesslike analysis by Capt. Augustus W. Corliss (1837–1908), is included. Corliss, a brevetted cavalry lieutenant

---

[1] No one seems exactly sure of how many wagons were involved, but a number in the two hundred range is likely.
[2] Edwin F. Townsend (1833–1909), West Point '54, spent most of the war in rear echelon administrative positions.

colonel during the war, rejoined the army as an enlistee in 1866, was a lieutenant with the Eighth Infantry in 1873, and was promoted to brigadier general on his retirement in 1901. Exactly why and to whom Corliss submitted his account of the survey is unknown, but it is straightforward enough and often insightful.

Barrows to the Tribune

THE YELLOWSTONE EXPEDITION.

A FORMIDABLE MARTIAL ARRAY—THE PROBLEM OF SUBSISTENCE—AN IMPRESSIVE CAVALRY REVIEW—GENERAL ORDERS.

FORT RICE, DAKOTA TERRITORY, JUNE 18.—"This looks like old times," said a cavalry officer to me as we stood on a high hill above his camp and cast our eyes over the tented field before us. "This reminds me of Washington and '61." The scene was indeed suggestive of the scenes and experiences of ten years ago. Immediately before us, on a lower shelf of the bluff, stood Fort Rice, with its high stockade and its garrison buildings, the only intrusion of modern architecture on the vast extent of rolling plain, save the sutler's store[3] and a few log houses built outside of its walls. On the right of the fort near the river, on another swell in the plain, a hundred pretty white tents dotted the green. The welcome dinner proclamation, the smoke of the camp-fire, curled into the air. Two or three hundred blue-coats lay scattered up and down through the avenues of white. Further to the left, on the bluff near the river, perhaps 50 more tents stood in array. Midway between the two on another piece of high ground was the commandant's tent with the flag flying from its top, and the tents of his staff clustering around it. Two miles away from headquarters, on a hill stretching a mile to the left from the base of the fort was the cavalry camp, like a city of white set on a hill. In a long line 200 horses were trotting down to the river to drink. Four hundred more were picketed around the camp. Down on the plain a drove of several hundred cattle were grazing. With the field glass might be seen the train of 250 wagons on the opposite side of the river, awaiting ferriage across.

Outside of the camp line on another hill, two [3" Rodman rifled] fieldpieces were playing ball with a target at a mile distant, now and then sending a shot across the river to tear up the dust on the eastern hills. There was no sign of civil life in the picture. The coloring was altogether martial. I have only to look from my tent to revive the picture. I presume the Yellowstone Expedition, though protective and peaceful rather than aggressive in its character, is the largest military expedition since the war;

---

[3]Sutlers sold provisions, tobacco, dry goods, and sometimes liquor.

and if the war had not changed our ideas of martial magnitude, would have been regarded as formidable indeed. For the purpose for which it is intended it is formidable. It will be a pity if the Indians cannot be convinced of this except by the voice of musketry and cannon.

THE LANDSCAPE ON THE UPPER MISSOURI.

Apart from the display of camp equipage, the flying horsemen, the round of the sentries, the call of the bugles, and the sound of the artillery, the scene is as peaceful and quiet as could be desired. From the height at which our camp stands, a green alternation of hill and plain rolls clear to the circle of sky. For miles in every direction the eye may wander over the waving green carpet without restraint. Yet the picture has no monotony. The blue horizon is scalloped by the impinging outline of hills. The whole surface of the landscape is grooved and wavy as if some mighty plow or harrow had furrowed the prairie into vale and intervale. Only on the east is the picture interrupted by the winding waters of the Missouri, swiftly moving to the south. There is a fringe of timber along the soiled skirts of the turbid river. Here and there an arm is stretched out to claim its earthy tribute from the hills. Except this tall edge of timber by the river, there is no sign of tree or shrub. Sage and snake-root and short, stunted grass are the principal feeders on this dusty, alkali soil, if we except the cattle, for whose benefit alone they seem to grow. From 50 to 75 miles or more back to the east, along the Northern Pacific road, the land is nearly all of this character. In a month or two it will be dry and brown under the scorching sun. Stock may grow and live here; but farmers will not find it inviting. Irrigation might do for this section what, through the labors of the Mormons, it has done in Utah, but the soil does not take kindly to growth. It is not the parent kind which was considered "good" in the Garden of Eden. Adam and Eve would not have fallen here, for the reason that all fruit is [foreboding] or forbidden. The beasts of the field may thrive and prosper, nibbling the dwarfed grass, for, if it does not grow high, it grows very widely. There is no lack of room. A few early vegetables and a variety of Indian maize are capable of being domesticated on these dry prairies, but wheat and other grains are not successful.

Preparations for the departure of the expedition are nearly completed. The wagon trains are crossing the river to-day, and will immediately be loaded. In two days the line will be formed and the march begun. Officers and men are getting impatient of delay. This has been occasioned by the late arrival of the wagon train. As in civil so in military life, one delay brings on another. The delay here, however, does not at all retard the expedition as a whole for the detachment sent in advance, Friday, enables the survey to begin immediately.

TRANSPORTATION AND SUBSISTENCE FOR AN ARMY.

Few except those who have been in army life know what extensive preparations must be made for the transportation and subsistence of 2,000 troops in a new country. In a settled region traversed by railroads or navigated by boats, transportation and supply are easy matters. Equipage and provisions and even troops themselves can be readily concentrated when desired. But in a march of 400 or 500 miles through a new and hostile region, removed from any base of supplies, it becomes a more serious matter. The troop must carry its own *impedimenta*, its own subsistence. This requires a large number of wagons and with not less than four or six mules to a wagon; this in turn needs a large number of teams. In addition to the rations and supplies for soldiers, forage must be carried for horses and mules. Where there is a large body of cavalry, as on this expedition, the forage-train must be increased. To insure a supply of fresh beef, a large drove of cattle on the hoof must be taken along. It will be seen that it is no small matter to provide for such a force. This large train of men and animals must be absolutely self-supporting. There are no towns or cities on the way where provisions and horses can be levied. There are no telegraphs to convey wants or railroads to supply them. The army must be complete and independent in itself; able to go off on four legs without limping in any one of them. An army of this kind may come back with crutches, but it will never serve any good purpose to send it off in that condition.

To give some idea of the preparation required for such an expedition, let me give a few figures:

Number of men, including teamsters, herders, and civilians of all sorts,	1,900.
Rations for 60 days,	114,000.
Bacon, lbs	23,428
Beef, lbs	101,787
Flour, lbs	42,750
Hard bread, lbs	76,000
Beans, lbs	8,550

Rice, lbs	8,550
Roast coffee, lbs	9,120
Sugar, lbs	17,100
Vinegar, galls	570
Candles, lbs	712
Soap, lbs	4,560
Salt, lbs	4,275
Pepper, lbs	285
Tobacco, lbs	3,800

This is the provision for but two months. At the end of that time the expedition expects to be at the Yellowstone River and receive supplies for two months more, which will just double the figures above.

There are, besides, 745 horses and 1,593 mules. To supply these animals, over 12,000 pounds of grain per day are necessary, or about 1,500,000 pounds for the trip. To transport this, with the necessary hospital stores and the appliances for the scientific party, requires 253 army wagons, 6 ambulances, 2 spring wagons, and 1 hospital wagon. Even this large train cannot transport the forage above-mentioned. It will be necessary to send some of it up by boat [to Glendive Creek], and a train will probably be sent back for more supplies after the expedition has gone a hundred miles.

THE CAVALRY—AN UNUSUALLY FINE REVIEW.

Yesterday, at the request of Gen. Custer, the 7th Cavalry was reviewed by Gen. Terry, Commander of the Department of Dakota. The regiment, numbering about 630 men, formed on the plain, near the fort, and went through a variety of pleasing evolutions embraced in the cavalry practice in a manner so satisfactory that Gen. Terry will officially congratulate the regiment. The officers, in their bright new uniforms with plumed helmets, made a very showy appearance, and the whole command—horses as well as men—won a verdict of praise which was well deserved. I have never anywhere seen such a fine company of horses as those ridden by the 7th Cavalry. They were mostly obtained in Kentucky. Some of them are thoroughbreds; all have been so well trained and are so thoroughly under control that it seemed as if they comprehended every order from the commander as quickly as the rider. The officers generally are fine riders and the men have had the advantage of considerable drill. For a dress parade nothing could be finer; for war of course there is another criterion. This regiment, however, is not merely of the dress parade sort. It was organized at the close of the war and has seen considerable Indian service. It contains many men who served through the war, and some of them participated in those dashing raids which gained for Gen. Custer his present reputation. The cavalry forms a prominent feature of the military combination, and should there be any Indian fighting will no doubt acquit itself creditably. With infantry and cavalry united this expedition seems strong enough to fight all the Indians in the country.

As the expedition will leave in a day or two, Gen. Stanley has issued the following order, which will govern the march:

HEADQUARTERS, YELLOWSTONE EXPEDITION,
FORT RICE, D.T., JUNE 15, 1873

GENERAL ORDERS NO. 4.—The following shall be order of calls, march, and encampment upon this expedition, to be modified as circumstances may require.

Reveille—Daybreak, men turning out under arms.
Breakfast Call—One half-hour after reveille.
General—Two hours after reveille.
Assembly—Two hours and fifteen minutes after reveille.
Advance—Two hours and twenty-five minutes after reveille.
Surgeon's Call—One hour before sunset.
Guard Mounting—One half-hour before sunset.
Retreat—Sunset.
Tattoo—9 o'clock P.M.

All calls will be sounded from these headquarters (except such other calls as are required by cavalry), and will be repeated by battalion commanders.

The infantry will alternate front and rear on the march [under the command of] Lieut. Col. L. P. Bradley [and] ... Major R. E. A. Crofton. ...

The cavalry will march on the flanks, five companies on each side of the train; the two Rodman guns will march at the head of the column.

Company "E," 22d Infantry, with the exception of those men detailed as artillery, will serve as pioneers.

The guard will be detailed by companies, and will be varied in strength, according to circumstances.

No straggling will be tolerated, and no person will be permitted to leave the column without the permission of the Colonel commanding the expedition.

Battalion commanders are authorized to permit selected men to fire at game coming near the column.

No person connected with this expedition will fire upon an Indian unless the latter should first fire or give unmistakable signs of being hostile.

Officers of infantry who furnish their own horses and equipment will be permitted to ride, but no forage will be allowed such animals.

No person will be allowed to ride upon a commissary wagon.

All animals put out to graze will be picketed or hobbled.

In addition to the regular guard, there will be a battalion of one non-commissioned officer and three privates daily, under charge of the field officers of the day.

The cavalry will perform their own guard duty.

The officer commanding the detachment of Indian scouts will report each morning to the Colonel commanding the expedition for orders.[4]

The infantry will encamp on the four sides of the train, and the cavalry by themselves, on the most suitable ground, at the head of the column.

Upon arriving in camp, and as soon as the battalion is in line, a picket from the three infantry battalions, the strength of which will be changed according to circumstances, will be sent to his proper front by each battalion commander, to the first high or commanding point, if not more than one-fourth of a mile distant. One man will be kept on watch at each outpost; the remainder of the picket may ground arms and rest. This picket may be relieved by an equal number of men from the same battalion, by order of the Battalion Commander, to enable the men to get meals. While on picket men will not take off their belts, and must keep together. The pickets will be relieved by the regular guard at sunset, and will resume their stations next morning upon the guard being relieved at sunrise. The pickets will rejoin their battalions when the advance is sounded. An officer or a non-commissioned officer detailed for picket must remain on strict watch while the picket is on duty. The field officer of the day is charged with the responsibility of observing the manner in which picket duty is done, and of correcting and reporting all delinquencies. Until further orders, the picket from each infantry battalion will consist of one commissioned officer and 15 enlisted men.

---

[4]Lt. Daniel H. Brush (1848–1920) enlisted in the army at sixteen, survived wartime combat, and graduated from West Point '71. He had similar responsibilities in 1872 and was highly respected by his scouts.

Rations will be issued every five days by the Chief Commissary of Subsistence, on consolidated battalion returns, and turned over in bulk to an officer designated by each Battalion Commander, who will issue to their respective companies. Scouts will be furnished each Battalion Commander. By order of
Col. D. S. Stanley
H. H. Ketchum, A.A.A. General.[5]

Published July 3, 1873

Bradley's Diary and Letter, June 13–19

JUNE 14TH. Col. Townsend left this morning [by steamer] with four companies of the 8th for Fort Lincoln to escort the engineers for the Northern Pacific Railroad out on the Heart River route in advance of the marching of the Expedition. Wagon train arrived today. All the material for the Expedition is on the ground now and we commenced loading.

LETTER NO. 11

In Camp
June 15, 1873

My Dear Love:
 . . . This is the day we were to have started out from here, but it looks as though we should not get off before the 20th. A large train of 250 teams came in yesterday from the east and will commence loading at once. One steamboat loaded with stores is yet to arrive and there I hope there will be no further detention. We might just as well been at home two weeks longer than not.

Col. Townsend started yesterday morning . . . and we shall join him 40 or 50 miles from here. I am tired of laying in camp to be eat up by mosquitoes and shall strike tents gladly any time.

We are all as well as possible and very comfortable too with the exception of the mosquitoes. I don't need anything more than I already have

---

[5]Hiram H. Ketchum (1843–98) was a Civil War veteran and acting adjutant general for Stanley in 1872 and 1873.

except my bathing sponge and am sorry I did not bring it. . . . Our Steamers for the Yellowstone are loading at Bismarck at the crossing of the Missouri will start soon. . . .

[Personal comments]

JUNE 16TH. Very hot weather and mosquitoes tormenting us day and night. Went over to see the review of the 7th Cavalry at sunset.

JUNE 17TH. Gen. Terry left today. . . . Mercury 102° in the shade.

JUNE 18TH. Heavy wind storm blew down the front of the tents. Wind changed into the north and was cold.

JUNE 19TH. High winds and cool day. Drilling every day principally in "Skirmish." Orders for marching.

Rosser's Diary and Letters, June 13–19

JUNE 13TH. Slide in track near Fond du Lac detained us 1½ hours, but made it up before getting to Brainerd. . . . Reached Fargo 8 P.M. Left for Bismarck 9 P.M.

JUNE 14TH. Reached Bismarck 10 A.M. Found I was quite ahead of time and could have spent another day with my precious family. Went to work on the estimate and very busy all day and until late at night.

JUNE 15TH. Finished estimate 6 P.M. and gave orders for work to begin tomorrow. Very warm and not feeling well, my suit is too hot and I fear I will suffer greatly from heat this summer.

JUNE 16TH. Steamer behind time and I took a life boat from steamer and went down to camp and got everything ready to commence work in A.M. The boat came down and with the engineering party and one company of infantry. We ran up river and commenced work. Ran [surveyed] 2½ miles. . . .

JUNE 17TH. Left camp this A.M. at 5 o'clock and when about one mile from Fort Lincoln ran into a squad of 50 hooting Indians but escaped without injury.[6] Ran three miles and walked five miles to camp.

---

[6]News of the fighting was telegraphed around the country, and one assumes Rosser wrote these lines believing his wife would eventually see them.

Heart River, D.T. [A.M.]
June 18

My Precious Lizzy;

I sent a telegram back by scout this morning to you which I hope will allay all your fears concerning us. As I was going out yesterday morning, Meigs and myself riding in front, I discovered an Indian not more than a hundred yards from me. And I ran back to get the escort and tried to deploy them to the point so as to be ready for the savages when they should rush out. But they were so slow that the Indians rushed out at us before the infantry got into place, but a few shots . . . were sufficient to drive them back and we turned over the fight to the Ree Indians and we went on to work. . . .

Kiss the dear children and your Mama for me, and please don't worry yourself sick on my account for I will take good care of myself and also of the party. I don't allow a member of the party to go outside of the line of sentinels. . . .

North Fork of Heart River [Sweet Briar Creek]
June 18th [P.M.]

I send you two letters today. . . . We have had a very quiet day of it today and are getting on very nicely, everybody enjoying themselves very much and the Indians keeping out of the way. The escort is entirely too large and looks like an immense army—sufficient to whip all the Indians in this territory. But oh, how disgusted are the officers to think they have been brought out in such force this summer.

My Bug Catcher (Prof. Phelps) as the boys call him, is getting tired of it already and I think when we reach the Yellowstone all of them will return east, and leave the expedition pretty bare. Mr. Roberts writes [*illegible*]. He expected me to remain with the expedition, but I now most solemnly promise you that I will not go beyond the Yellowstone if there is a shadow of a chance to get back. And as we cannot cross the Yellowstone without a boat, and as that boat has to return down the river, I think I have a perfectly sure thing on getting home early in August. There is no doubt of our getting to the crossing of the Yellowstone by the 1st of August, and it will not take many days to run down the stream when I get started.

[Personal remarks]

JUNE 18TH. Moved camp to the Big Bend of the Heart River where the line reached us about 5 P.M. Sent messenger to Bismarck with letters and telegrams. Hope to hear from dear Lizzy by return messenger.

JUNE 19TH. At work up the Sweet Briar Creek. Stream very crooked but pretty valley and some timber. Water good and abundant. Several fresh signs of Indians but saw no sign of Indians. Boys had a long walk to camp.

The following appeared on the first page, upper left hand corner, of the June 18 St. Paul [Daily] Pioneer and numerous other papers.

SPECIAL DISPATCH TO THE *ST. PAUL PIONEER*.

BISMARCK, D.T., JUNE 17.—The Yellowstone Expedition, which left Fort Lincoln this morning, were attacked by about 150 Sioux about two miles West of Fort Lincoln, the result of which was four Sioux killed and one Ree scout slightly injured. Nobody else hurt. It was a species of bravado on the part of the Sioux, and has no significance whatever. There is no probability of any further attack or raid on the part of the Indians until Powder River country is reached and the expedition is so formidable that no open attack can be reasonably expected.

Report of Augustus W. Corliss
THE YELLOWSTONE EXPEDITION OF 1873
TROOPS

The forces employed in this expedition . . . were thoroughly equipped for field service, armed with breech loading rifles of the best pattern then in use, the Springfield, Cal. 50, the new brass-screwed shoes, campaign hats, etc. No bayonets were carried as they are useless in an Indian campaign. The knapsacks were always carried in the company wagons. The men carried their rifles, forty rounds of cartridges, canteen, and haversacks.

ENGINEERS

The surveyors employed by the railroad were all citizens, about thirty in number, all under charge of Gen. Thomas L. Rosser, of Confederate fame, a classmate of Gen. Custer, and were supplied with the latest and best instruments obtainable. Their train of wagons was most admirably fitted for their use and were wonders of labor-saving and space-economizing expedients, most of them the result of Gen. Rosser's experience during the war. Their tents were large, specially made affairs, fitted up with folding chairs, tables, drawing stands, etc., for the use of the engineers in making up their maps after reaching camp; and a thorough system prevailed in every part of the whole organization. Duplicate notes of the surveys were made, and one complete set kept in the tent of the commander of the troops so that in case of accident by fire or flood all would not be lost. The cooking of the engineer party was so much superior to that of the messes of the officers of the troops, that a detail as part of the daily escort was always looked upon as a picnic as the officers were always invited to luncheon with Gen. Rosser.

Every day a detail of from four or more companies of infantry was made as engineers' escort, and were usually deployed as skirmishes on each flank of the workmen and far enough away to keep an attacking force off until the surveyors and chainmen could rally under some good cover. Affairs of this kind occurred several times, and the alacrity of the men in hunting holes for themselves and their instruments was something wonderful. Before starting out in the morning a plan was agreed upon for the march of the day, and then the main body with the train would cut loose from the surveyors for the day.

One day this plan miscarried, and the engineers and their escort remained out all night without food or shelter, except that a grizzly bear and cubs furnished small rations to the men.

THE [WAGON] TRAIN

The train, excluding that of the engineers, was composed of about three hundred army wagons, each drawn by six mules, and about a dozen ambulances and spring wagons. These were divided into sections, as follows: forage, rations and sustenance stores for officers, ammunition, repairing material and wheelwright tools and parts of wagons and gun carriages. All were appropriately marked, and each section had its place in the train. One wagon, with tools and bridge materials for the pioneers, was kept at the head of the column. When a wagon broke down it was pulled out of the column, and the wheelwright section drove to it at full speed, and in a few minutes the broken wheel or tongue was replaced, and the place in the column [re]gained at the next rest.

To each company two wagons were assigned, and in these were carried five days rations, tents, ammunition, baggage, knapsacks, etc. The company wagons were at the rear of the train. The wagons moved in one, two or four columns, according to the character of the country. There were no roads. At any alarm [Indians], the wagons, if on the move, were driven at a gallop to form a "corral," those on the right driving up to those in front, which had halted with the mules of each wagon sheltered behind the left side of the next wagon in front of it. The left columns formed by inverse means, so that all the animals were out of fire. The animals of the leading wagons were taken inside the train. Ropes were tied across the front opening of the train. The beef herd and spare animals were run into the corral at once.

When camp was formed, the wagons were parked by sections. The wheelwrights and blacksmiths went to work to repair and make good the accidents of the day. The subsistence and forage trains began to issue rations, officers stores and forage to all in need of them. The system was perfect and no time was lost and no confusion was ever seen.

As wagons became empty, the loads of others were equalized and the picket-guard[s] of the previous night were allowed certain wagons to sleep in during the day's march. Empty casks were carried on all wagons, and when any uncertainty was felt about water ahead for the next camp, they were filled before starting in the morning. At night the mules were hitched to the wagons, and a corral guard from the drivers watched the whole train and animals.

CAMPS

Camps, of course, were made with reference to facilities for wood, water and defense in case of attack. As a rule, then, they were located on the mesa near a creek, the wagons parked, the troops camped in a huge circle outside the train, from seventy-five to one hundred yards from the wagons, the tents of the men, facing outward, covering the entire train, those of the officers just back of the men's line. The picket-guard, made by entire companies for each flank, front and rear, was posted just outside the line of the men's tents, with sentries about fifty yards in front. In case of alarm, the entire guard fell back inside the lines. Every man moved out about ten yards in front of his tent and lay down in line to await orders. The officers lay down behind the men, and in one minute from the first alarm over two thousand rifles and carbines, in the hands of cool men, were ready for business. The camp and kitchen fires were in front. At Taps they were extinguished. One hour before daybreak all stood to arms and the picket-guard was pushed out well to the front. The cattle herd was kept at the corral. Before breaking camp, all fires were extinguished.

MARCHES

... [T]he men stood to arms an hour before daybreak. As soon as it was fairly light, tents were struck, wagons packed and preparations made to march. The pickets were withdrawn and Indian scouts explored the country on every side for any hidden foe waiting for us. As soon as Gen. Stanley got into his saddle, one-half of the cavalry moved to the front about a mile and halted. The other half moved to the rear a short distance and waited for the train to move out.

The train pulled out in two or four columns. The infantry battalions, four in number, moved in columns of fours, one in front of the train, one on each flank and one in [the] rear, and from fifty to seventy-five from the wagons. When from any cause the train stopped, the infantry halted. The Indian scouts hovered on all sides, from one to two miles away, and explored every place where a foe could be concealed. In case the scouts came in on a gallop, firing their rifles, the train was corralled at a dead run, and the infantry threw out clouds of skirmishers at double time and moved out to prevent any shots from reaching the train. The cattle herd moved in rear of the train and, in rear of them, one-half of the cavalry. In case of an alarm, the herd was moved rapidly to the corral. At the end of every hour the troops halted for a rest of ten minutes,[7] and the train closed up solidly. The infantry battalions rotated daily from front to left flank, rear and right flank, and each leading company in each battalion took the rear of the column the next day. The men usually put their blouses in the wagons at starting, but carried their rifles, cartridge-belts, haversacks and canteens.

The distances marched in a day varied according to the work of the engineers, and, of course, to reach water, but, as a rule, averaged from fifteen to twenty miles each day. The most ever made was thirty-two miles,[8] but that was an emergency to reach water. . . .

---

[7]This break was ridiculed by cavalry officers, but eighty-nine years later this editor, actually young at one point, learned and will attest that such breaks *are* necessary for marching infantry.

[8]After leaving the Musselshell in order to reach Big Porcupine Creek.

At about 11 o'clock A.M. a halt for one-half an hour was made and officers and men sat down for a luncheon and smoke. This course was kept up day after day and month after month. The infantry, after the first few days, were as hardy as men could be, and seldom was a man on sick report. The cavalry horses suffered somewhat from sore backs, and forage reduced one-half but, as a rule, were in good condition. As soon as camp was formed, guards were posted and all the animals were turned out to graze. Every hilltop was occupied by an Indian scout, and every commanding ridge by an infantry picket-guard. . . .

[One day we marched] over the bed of an old sea, a mass of fossils and petrifactions, but we were more interested in finding *water* than we were in the remains of extinct animals and fish, and plodded, weary and hot, over what would have excited our greatest curiosity at any other time.

Source: Indian Wars, Miscellaneous Collections folder, U.S. Army Military History Institute, Carlisle, Pa. Corliss's handwritten report ends abruptly on page 22, with an unknown number of pages missing.

PART II

To the Yellowstone

CHAPTER 4

June 20 through July 7

To the relief of all, Stanley's column marched west from Fort Rice on June 20. In contrast to the attack on the surveyors three days earlier, a handful of Sioux let themselves be seen but made no effort to impede the three-mile long column. Nevertheless, the army soon faced another problem: incessant rain.

With one downpour following another, the Plains turned into a sea of mud, far too soggy to support the three-ton weight of most wagons.[1] Days later, the surveyors' column was overwhelmed by a vicious hailstorm that panicked the horses and mules and led to the damage or destruction of all its wagons. Finally Stanley found himself blocked by a usually placid stream (Big Muddy Creek) that swept away all his bridging equipment. These delays and lack of sleep, beginning the first night out, spared no one: rare was the individual who was not soaked or had his tent blown over. Soon everyone had become edgy. Worse, it magnified the normal animosity between infantry and cavalry officers.

However, the march was not without humor. For example, the soldiers were initially allowed to shoot at nearby game, and Barrows notes Stanley's comment that the safest place to stand was directly behind an antelope.

Chapter 4 begins with Barrows's *New York Tribune* letter published on July 25. Following it Rosser and Meigs describe being caught in the hailstorm. Also included are Bradley and Konopicky's diaries and letters and the *Sioux City Journal*'s account of the army's snafu-prone supply steamboats coming up the Yellowstone.

---

[1] Besides carrying some 5,200 pounds of supplies, the wagons themselves were made of heavy wood weighing hundreds of pounds.

Barrows to the Tribune

**YELLOWSTONE EXPEDITION:
EXPERIENCE ON THE WESTERN PLAINS.**

**A REVEILLE THAT WAS NOT UNWELCOME—
THE POWER OF MUSIC—AN EXTRAORDINARY SERIES
OF THUNDER-STORMS—MEN AND ANIMALS
MUCH BRUISED AND INJURED BY HAIL—
SCIENTIFIC DISCOVERIES.**

FORT RICE, DAKOTA, JUNE 20.—If on some fine Summer morning, at the early hour of 3, when the hot, scorching sun of midday was taking its morning bath in the Atlantic, and the air was cool and sleep-favoring, a man with a bugle should come to your front door in New York and set up a brassy, blatant staccato toot to mar your pleasant dreams, would you not have a rare ear for music if you did not think of boot-jacks, blacking-brushes, and other winged instruments of domestic use as a suitable response? It is no affected virtue which prompts me to say, however, that the bugle, at that hour this morning at my front door, had no such effect. Its sound was welcome indeed. Its notes were literally stirring. They had the same effect on this army that they would have on you or your neighbor if continued long enough. They caused us to change our location, and the change was welcome. Officers and men were glad to get off. Even the horses and mules needed some fresh grass for their picket. Fort Rice and environs had already become familiar. It bred no contempt, indeed; but nearly every one was willing to say "Good-by."

In a few minutes after reveille every one was in his clothes, with his duds all packed, for the reveille was imperative. However offensive the bugle may be at some times, its authority is unlimited. There are no newspapers. . . . The bugle is the only way they have of publishing the laws. So if the reveille did not bring you on your feet and into your clothes on this marching morning, the strikers would very soon have the tent down over your head. I know a General who, on one occasion when his officer failed to obey the reveille call, ordered his tent to be fired immediately. The order was obeyed. I think water would have been more economical and quite as efficient.

THE EXPEDITION IN MOTION.

The camp soon presented a lively appearance. One after another the tents came down and were quickly transferred to the quartermaster's wagons. Piles of baggage and bedding quickly followed them. Before the break of day the cooks had lighted their fires. Just as the eastern clouds had begun to blush at the first advances of the morn we sat down to a breakfast of bacon, potato-balls, coffee, and biscuit. It is one of the mitigations of the bugle that it announces breakfast and rest as well as reveille and advance. The breakfast is eaten. The wagons are all loaded. The soldiers are ready. The "advance" is sounded. The infantry form in line, the wagons in train, the cavalry mount, the band strikes a tune, and the Yellowstone Expedition moves on its way.

It is one necessity of this nomadic life we have undertaken that house moving is reduced to a fine art. . . . Moving becomes a privilege. It is part of our everyday business. I fancy I see some private soldier, perhaps an officer—if officers did such things—point over his left shoulder when I speak of the "privilege" of moving. But such doubtless are not real soldiers. They only wear the soldier's uniform. It is the soldier's privilege to march quite as much as it is to garrison. Change is the stimulus that relieves the fatigue. Omitting severe and unusual marches, there are few soldiers of any sort who would not rather, during pleasant months of the year, pack and unpack and march eight or ten or more miles every day, than loaf around a post.

APPEARANCE OF THE FORCE.

"Let us gallop to the top of that hill," said Gen. Stanley, "and take a look at the column." We spurred our horses, and soon reached the crest. Here we had a full view of the moving panorama, as it wound its serpentine length over the knolls and undulations of the prairie. A company of Indian scouts, 30 in number, under the command of Lieut. Brush, headed the line. About 20 yards in advance of the scouts rides Clemmo, the chief scout and guide.[2] He it is that we expect to pilot this expedition clear to the Yellowstone, about 290 miles. Beyond that point he carries no maps in his brain, and we shall look to heaven for guidance. . . . The

---

[2] Basil Clement or Claymore (1824–1910) was of French Canadian background. "Clemmo" was not a nickname as much as a mispronunciation of his surname.

faithful compass shall prescribe our path. But the stars and the compass and the surveyors' instruments will not tell us where wood and water can be found. We shall still need Clemmo and his scouts not a little.

Behind the scouts are the two pieces of artillery. Then a detachment of infantry. Then comes the long train of wagons and ambulances amounting to [*line missing*] Five companies of cavalry guard each flank of the train. The herd of cattle follow, and a detachment of infantry close up the line. One of the prettiest features of the picture is the long wavy line of wagons with new white bow covers, seeming newer and whiter under the freshening rays of the sun. For three miles the line stretches over the plain. For the first mile the outline of wagons and mules is distinct enough; but with the receding perspective it grows smaller and smaller until I can compare the diminuendo to nothing else than a train of old-fashioned coal-scuttle bonnets drawn by a six-rat team. The long, neat line of cavalry becomes so many elfin horsemen, whom daylight has caught on the hills. Soon the cavalry band strikes up a familiar air to add to the "pomp and circumstance of war." It was a good idea of Gen. Custer to take the band. It is not the finest band in existence, to be sure. It would hardly claim a place in Thomas's orchestra.[3] It is a band that has been with Custer and his cavalry for some time, and is used to roughing it. The music may be a little rough, too, for want of the smoothing effect of sufficient practice. But whatever its deficiencies, its music is welcome enough. Music and motion are sisters. It is singular what sympathy they have for each other. Every officer knows how much a band can do to revive the wilted spirits of soldiers on a long and tiresome march. It has a tonic effect on the mind, which soothes the nerves and strengthens the muscles. It needs, of course, a receptive ear. The man who is not thus provided lacks a great source of strength. In his Winter campaign against the Indians, in '68 I think, Gen. Custer, with his men, suffered great hardship. There was a sad and irreparable lack of rations. All that Gen. Custer could do in this extremity was to feed his men through the ears instead of through their mouths. There was a certain ration of band for breakfast, dinner and supper, with liberal lunches through the day. Perhaps there was a certain sameness in the quality, which induced one of the soldiers to complain that Gen. Custer fed them on "one piece of hard tack a day and the 'Arkansas Traveler.'"

---

[3] Theodore Thomas (1835–1905) was a well-known maestro who founded the Chicago Symphony Orchestra.

It was not deemed fit on the first day to make a long march. With such a large number of men and animals, wood and water near the camping grounds are of great importance. The length of marches may thus vary considerably, according to the interval between these camp necessities. A place called Eleven-mile Creek . . . was fixed upon for the first day's halt. Beyond that point wood and water could not be obtained without too long a march. A short thunder storm in the morning dampened no spirits, but brought much refreshment. The trail we followed was the one taken last year to the Yellowstone River. It leads directly west through a corrugated country tolerably well watered, but with very little timber. A few miles on this road, west of Fort Rice, it is barren enough. Sandy buttes arise here and there, and the ground in large patches is hard, parched, cracked, and unfruitful. But further on there is a rolling country of tall, nutritious grass, which would make a perfect paradise for cattle. . . .

The absence of any late rains had left the trail comparatively dry, and the first day's march was performed without great difficulty. Two or three deep gullies gave some trouble to our heavily loaded teams. Each of the commissary and forage teams is loaded up to the top of the wagon bow, and carries a weight of not less than 5,200 pounds. This weight is drawn by six mules. In two places the pioneers[4] were called to reduce the steepness of the banks and cover the muddy bottom with sod, dry earth, and grass. The tall grass cut by the mowers heaped plentifully over the marshy surface served as a corduroy and kept the wheels from sinking.

CAMPING AT ELEVEN-MILE CREEK.

By 11 o'clock in the morning, the head of the column had reached the camping-ground. It was three hours before the last of the wagons came into the corral. Here we found good water and enough dry wood to cook our dinners. It was a great relief to find the water, though somewhat alkaline, entirely free from mud. For a week before leaving Fort Rice I had been drinking mud gruel dipped from the Missouri. The river comes honestly by its name of "Big Muddy." Its water is excellent, if filtered; but the amount of sediment it contains at this season of the year is extraordinary. Take a common tin wash-basin, fill it with water, then scoop up a heavy handful of mud and stir it in, and you can have

---

[4]Infantrymen who built temporary roads and bridges.

some idea of the color and character of the Missouri. This is the water that people on its banks use for washing and drinking. If a quantity is barreled and allowed to stand, the sediment soon settles and the water becomes tolerably clear. At our camp condensed milk was used to settle it with great effect. A small quantity placed in a glass of water precipitates the mud almost immediately. I have many times had to take it both for drinking and washing directly from the river. I have discovered no deleterious effects from drinking the muddy water, but often in washing in it have questioned whether I should not be less clean after washing than I was before. Another pest I hope we have left behind us is the Missouri musketo. At Fort Rice musketoes were not unmercifully abundant; but further up the river, at Forts Lincoln and Bismarck, they were almost unbearable. The Indians do not give one-half of the trouble. It is only by wearing a musketo shield over the head and gloves on the hands that protection is secured. . . .

In six or eight days we expect to make a junction with the surveyors, who are now running their line along the Heart River under the protection of the infantry and the cavalry company detached last week. When the junction is made it is not expected that the party will again be divided.

A THUNDER-STORM ON THE PLAINS.

CAMP NO. 5, JUNE 26.—A writer in the East,[5] identical with the man who proposes to make Russia bankrupt through the Hoosac Tunnel, has written somewhat about Dakota. He is remembered here mainly for one fact. He is charged with saying that it never rains here. His stay in the Territory must have been short and fortunate for his comfort. If our preparation had been based on his authority, we should have had a sorry experience. Indeed, it required all the resources we possessed for the first three nights to fortify our cheer against the drench and the gloom. Two of the heaviest thunder-storms I have ever witnessed I have seen in Dakota. The rainfall was very great, the thunder and lightning fearful, and the wind a perfect tornado. The felicity of camp life was assailed the very first night of this tempestuous quartette of thunder and lightning, wind and rain. Sleep was arrested, tents blown down, bedding flooded, and our night's rest completely broken up. The only consolation for this

---

[5]Barrows's reference is unclear. The 4¾-mile Hoosac Railroad Tunnel, finished in 1875 near North Adams, Mass., was a shortcut connecting Albany and Boston.

very bad night was that the next one might be better. The premise was hopeful; the conclusion false. There was only one night that could be worse than the first one, and that was the second. If all the artillery in the Ordnance Department could be parked, loaded, and fired together, the simultaneous roar would compare with that second night's thunder only as a popgun compares with a Rodman. I doubt if any possible gunpowder chorus could liberate a sound worthy of respect as a faint echo to a first-class Dakota thunder-peal.

Any man who could sleep through these successive crashes of ear-bursting noise must be mourably [sic] deaf. Flashes of lightning blinding in intensity and marvelous in rapidity rent the heaven with their zigzag knives. The wind became a hurricane flying across the plain. The rain poured in torrents. The thermometer dropped from a shadeless heat of about 120° to 47°. The thunder was bearable. The lightning only penetrated the swollen clouds and descended to no lower thrusts. Both could be accepted with resignation. Even the rain could be endured had it fallen with more rectitude of purpose. But the wind was the arch-disturber. It made cork screws of the rainfalls and twisted them under the tent flaps, soaking our blankets, and giving the blanketed [men] a wet-sheet pack. This was a hard test for patience and piety. A good many resolutions were broken. Interjections spontaneously came to the surface and were carried off on the swift air.

The storm was anything but moral in its influence. When at last the boisterous wind took many of the tents in its great strong arms and turned them over on their occupants, the cup was full. Only the angel who takes down naughty words, and who, it seems to me, might be better employed, could give an accurate report of the short and pointed speeches which this active wind excited. THE TRIBUNE establishment happily escaped humiliation; but not so the tent of our Chief Commissary Ray.[6] The wind unkindly leveled his tent poles and folded the canvas over him. He bore it with the indifference of a stoic and lay there till he was ready to dress. Two English gentlemen . . . , who at their own expense accompany the expedition for the purpose of hunting, suffered in the same way. One of these gentlemen is reported to have been seen at midnight on the prairie in the driving wind and rain, with an ax in

---

[6]Lt. Phillip H. Ray (1842–1910) enlisted during the Civil War and retired as a brigadier general after a career that included assignments against the Apaches, police duty in the Yukon, and fighting in Cuba and the Philippines.

his hands and a pair of moccasins on his feet. Otherwise he was totally destitute of vesture. It is said that as he drove the stakes into the ground he talked considerably to himself.

THE LONGEST DAY IN THE YEAR.

Our second day's march was about 16 or 18 miles. The heat was blistering in intensity. It was hard on our horses and wilting to the infantry. Several men dropped by the way and had to be taken into ambulances. No serious cases of sunstroke, however, are reported. The march was slow and the rests frequent. But on this treeless plain there was no possibility of shade. It was not until 3 o'clock in the afternoon that we camped by a little creek where there was abundant wood and shade. Never did water seem so precious or shade a greater boon. It was the longest day in the year, and I shall not forget it. The hours seemed to expand with the heat, and the heat to grow with the hours. Anaxagoras[7] sinned only very mildly when he called the sun a red-hot stone. He would never have been put to death on this expedition for such a gentle though unscientific observation. The sweetest portion of the day here is the sunset and the twilight that succeeds. It is a very small bar of darkness that divides the days in this Northern latitude. I can read now until 9 o'clock at night. It is not until 11 o'clock or later that the soft twilight shades into darkness. Then at 3 o'clock in the morning the first rays of daylight illumine the East, and the bugle sounds a call to a new day.

On our second day from Fort Rice we saw plenty of antelope. Gen. Custer with his dogs ran down several young fawn, and the hunters along the line brought down several with their rifles. But by far the most successful was the hunting club, composed of officers of the 22d regiment. They have several valuable hounds, chief of which is "Given." . . . Last year "Given" caught 12 full-grown antelope. This morning he was very successful. Before noon he ran down four antelope, which, considering the heat of the day and the remarkable speed of that animal, was a good day's work. The chase added some fresh, tender meat to our larder, and afforded great diversion for the men. When the dogs were let loose, the tedium of the march was forgotten. Shouts and yells ran along the line to cheer the dogs and confuse the antelope. The dogs were fired with enthusiasm themselves. The sight of an antelope 400 yards off was

---

[7] Fifth-century Greek philosopher (ca. 500–428 BC) persecuted in Athens for his scientific beliefs in explaining events of nature (versus those performed by the gods).

enough to start them when there was not a shadow of a chance of their catching him. To prevent them from running themselves out altogether it was necessary to leash them early in the day. Two spent dogs were placed in the ambulance. One unfortunate fox-hound of Gen. Custer's was accidentally shot by a hunter. . . .

CHASING THE ANTELOPE.

From the many knolls which lay in our path, commanding extensive views of the county, we could frequently watch a chase from the start to the close, and where the view was interrupted the temptation to follow at full speed with our horses was seldom resisted, when the issue was close. Indians were entirely forgotten in the rage for antelope. I cannot resist the impression that for purpose of sport alone it is wanton cruelty to hunt down this beautiful creature of the plain. But when one has lived on ham or bacon for several days, with only an occasional relief of tough beef, the impression changes. Antelope becomes a necessity. The sport is only a concomitant. It is a hard thing to remember when you are hunting antelope or other game that you are going to a funeral. If one succeeded in remembering this, an undue regard for conventional ceremony might defeat his purpose. The sport really lies in catching the fleet animal. It is only when you have reached him that you think of the knife. There is plenty of room for pity at death. And I do not think that among army officers this element is regarded as a weakness.

It is assumed when the old bucks are caught that they have lived long enough, and the time has come for them to serve their day and generation. But let the dogs run down some young fawn but a month or two old, and the father-element comes to the surface. I only tell the following incident . . . to show you that the men that command and compose the expedition are not all savages, even if they don't believe in the President's peace policy as a practical measure. I was riding yesterday with Gen. Stanley, Fred. Grant, and several of the General's staff at the head of the column just in advance of the party of scouts. As we walked our horses four abreast, we startled a mother grouse from the nest. She flew over our heads and settled a short distance off, within easy range. One of the party dismounted and found the nest in the trail. It was full of eggs just ready to hatch. "Poor things; let them be," said the commander, and the eggs were replaced, the rifles lowered, and the column of horsemen parted. . . . It may seem an unnecessary and trifling thing to turn an army from its path to spare a nest of eggs to a hovering bird, but anything that gives the better part of human nature a chance to come to the surface is worth the trouble it costs.

In accordance with the order of Gen. Stanley issued before the starting of the Expedition, a certain number of men along the line were allowed to shoot at antelope and other game. The General was desirous of giving the men every opportunity for hunting. But it was soon found that the liberty was greatly abused. The men in their excitement after game were not at all particular which way they fired so long as they aimed at an antelope. Frequently an antelope would pass between the hunter and the column. A ball whizzing over our heads showed us that the hunter appreciated his situation but did not appreciate ours. During the day balls came pouring in thick and fast from both flanks. I doubt if we shall receive many heavier volleys from the Indians than we received the second and third days of our march from the hunters. It is a wonder that no one was killed. An erratic ball came within a brief distance of my horse and struck still nearer to Dr. Kimball's orderly. Gen. Stanley was finally compelled to modify his previous order and prohibit any firing whatever except by his own special permission. This modification will effect a great saving in ammunition and will make no difference in the amount of antelope shot, for, as the General remarked on this wild shooting, the safest place for the last few days has been immediately behind an antelope.

SCIENTIFIC OBSERVATIONS.

Since leaving Fort Rice our scientists have not been idle. Mr. Allen has collected about 125 species of flowers between Fort Rice and Camp No. 5, a distance of 45 miles. The prevalent families are the Compositæ and Leguminosæ. The whole surface of the plain is colored with Yellow Asters and pretty little Purple Lupins. The most characteristic plants are the Astragelus Carjocarpus and the Amorpha Canadenis. There is another smaller species of Amorpha, exceedingly fragrant, not given in the botanies. It probably does not occur east of the Missouri. There are two species of Foxglove, the smaller one allied to the Digitalis of the East. Birds are somewhat scarce. There are not more than six or seven species that are at all common; not more than two or three that may be considered abundant on the Plains proper. Woodland birds are met in clumps of timber along the streams. The Kingbird especially is found in the bushes on the ravines, also the Arkansas Flycatcher and the Brewer's Blackbird. The birds of the Plains are the Meadow Lark and the Upland Plover. The Curlew is occasionally seen, and a few Grouse. The most conspicuous bird is the Chestnut colored Bunting. McCown's Bunting is occasionally seen. With these occur the yellow-winged and the bay-winged Sparrow; but they keep mostly in the grass and are not often seen.

The timber in this section is mostly cotton-wood; and much of it is dead. The green timber is mainly box and elder. There is some elm and a species of birch. Timber on the whole is very scarce and occurs only along the streams, which are not very common.

In the vicinity of Fort Rice and for 20 or 30 miles to the west of it over our route, a glacial drift, consisting mainly of quartzite and feldspathic [*sic*] rock, is scattered over the plain in great abundance. On the fifth and sixth days of our march this drift disappeared much to our personal comfort. It is not pleasant to camp on, especially if your bed is made right over the rough protruding surface of some quartz rock. The soil changed to a sandy loam, not very valuable for agricultural purposes, but suitable for grazing. The rocks consist mainly of clays and sands. On the 24th we saw beds of sandstone of 20 and 30 feet in thickness, scarcely indurated at all. In the lower beds of these sandstones Prof. Allen found strata of three and four feet in thickness, made up mostly of univalve shells.

On the 24th we reached the Dog-teeth Buttes, so called by the Indians from their peculiar conformation. From the summit of one of the tall, isolated buttes of this range, Heart Butte could be seen 20 miles away. . . .

To the casual observer, it is somewhat difficult to realize that these abrupt, insular hills bear on their caps what was once the surface soil of this whole region. But such is the testimony of geology. Denudation has worn the plain into hills and furrows, and, combined with the action of the wind, has left these conical buttes like islands rising from a green billowy sea. In some cases they form almost perfect cones, fifty, sixty, or seventy feet above the plain. In other cases they are jagged piles of rock, of great irregularity in form, sometimes tumbled together in broken piles, like the remains of an ancient ruin. On the butte we ascended, the outer wall of rock was elevated two feet above the interior, forming a circular bastion, a natural fortification. The rock was mainly sandstone of varying degrees of hardness, sometimes so soft as readily to crumble in the hand and be carried off by the swift wind, again very hard and argillaceous. Nodules of iron were found imbedded in the sand, and a dyke of trap was exposed on the crest. Wind and water had left their autographs with great legibility and in great abundance. Pockets in the sandstone showed how the wind with its siliceous augur had bored into the stone. Water-worn caves and sluices sometimes extended clear through the crest of the butte. Smooth, round holes were worn through the slabs of rock as clean as if made by a cannon-ball. Riding over the surface of the plain, we could look through these riddled bastions and see the sky beyond.

Our photographic artist [William Pywell] has been unfortunate. The spring wagon which was to transport his material failed to come to hand at Fort Rice. Gen. Stanley sent a detachment of cavalry to Fort Lincoln to look after it. They overtook us on the march with tidings that the wagon had gone to the Yellowstone River by boat, with the extra forage supplies. Unless he can improvise a portable gallery, we shall fail to preserve some of the finest views. Mr. Konopicky, our present artist, has, however, made some valuable sketches.

As yet we have seen no Indians. One of our pickets became scared one night, and after firing at some harmless shadow, called out the 22d Regiment. They marched up the hill and then marched down again, and probably slept better for the exercise. Two of our party who were out hunting elk struck a fresh Indian trail a few miles from camp; but concluded not to pursue it lest the pursuit might be reciprocated.

ONE NIGHT'S EXPERIENCE.

CAMP NO. 10, 50 MILES WEST OF FORT RICE, JULY 3.—For full nine days after leaving Fort Rice the persecution of rain and wind was continued. In a third very serious storm which I have not before chronicled, hail was added to the programme of destructive agents. I regret that no provision was made on this trip for recording the amount of rain-fall. But I do not think it would have made much difference with our comfort. In aggravated cases of bed sores, water beds are quite comfortable. . . . But the drift had disappeared before the third storm came on, and I could not see the beneficent provision in the two inches of water which it left in my tent. Rather than lie in a cold bath I dug a trench through the floor which, in a literal sense, was a green carpet of earth. A poncho and a buffalo robe served as another intervention. Somehow when the heavy wind came up it developed a strong distrust in the integrity of my tent-poles. They were, indeed, upright in their intention, but too easily influenced. Under the action of the wind they showed a strong inclination.

I got under the blankets and launched into a somewhat shallow sleep; no dream that "I dwelt in marble halls" could dispel the presentment. It was just as well. My tent would not support the dream and my dream would certainly never support the tent. Even the actualities of pins and ropes were unable to resist the fearful tornado which at midnight invaded the garrison. Sentinels and pickets were of no avail. Our ammunition was useless. In a prairie fire, the only way to combat it is by a counterfire. But this homeopathic principle does not apply to the winds; and if it did, we had only forty buglers and the cavalry band to resist it. What could we do with these against this tornado, backed up 2,000 miles to the east? The wind had victory in its breath. It charged through the camp and won the battle.

When about 3 o'clock the bugle sounded, a hundred lay on the tented field. This happens every morning; but on this occasion almost every man's tent lay on top of him. The wind had possession of the field. To call it wind is a euphemism. It was a full-grown hurricane. It was no respecter of persons. In its midnight raid on our camp no regard was paid to rank or station. Etiquette was past consideration. It blew over Gen. Stanley's tent with no more respect than if he had been a private soldier. One of his aides was served in the same way. Your correspondent could no longer claim exemption. A heavy blast . . . struck his triangular canvas on its windward leg and carried it off its feet as if it had been a toy set up for the mere purpose of being blown down. In one of the battalions nearly every tent was leveled to the ground. Happily the storm ceased by daybreak, and we were enabled to gather up our scattered effects on the plain and put ourselves together without further torment from the disjointing wind and the soaking rain. . . . To chronic grumblers the event supplies material for weeks to come. Unless one is disposed to take a cheerful view of such a misfortune, there is nothing which can excite more aggravation than to have your tent posts come rattling down on your head, while a white counterpane of cold, wet canvas is added to your bed. To get up at daylight, wring yourself dry, and make your toilet on the plain in full view of those who have been more fortunate is not the least part of the aggravation.

BAD WEATHER—DAKOTA MUD.

I have given you one night's experience in this nine days' battle with the elements. It is only a small segment of the circle. Day after day we were beset with embarrassments quite as aggravating. Gen. Stanley has been in this Department for some years; but he confesses that he has never seen the weather equaled. The rain has not been without intermission, but the intermissions have never reached 25 hours in length. Then the rain, when it has come, has come like a deluge. Fortunately we have had but one hail-storm. From this, the most dangerous of storms to an expedition of this kind, the main part of the command and outfit was happily relieved. The surveyors of the escort under Major Townsend that accompanied them, who were about 80 miles distant from us at the time, suffered severely. An account of the storm and the damage it effected is given further on.

So far, not excepting the Indian skirmish near Fort Lincoln, an insignificant affair, and the antelope hunters of our own party, who distinguished themselves by firing into our column, the only enemies of consequence we have encountered have been harmless thunder and lightning and the wind, hail, and rain. Apart from the disagreeable effects of marching for two or three hours at a spell in a drenching rain and then camping on a wet ground, this excessive moisture has given effect to a less impulsive but more insidious enemy, which for nine days has been constantly under our feet—an enemy which the more it is trodden on the more dangerous it becomes. I refer to Dakota mud. I do not believe that any body of Indians we are likely to meet could do us the evil that this malconsistent mud has done. To draw 5,200 pounds on a wooden pavement would not be a very heavy task for six mules; but to draw it through a heavy, sluggish, greedy mud, disposed to swallow everything it can get, is utterly exhausting to the strength and endurance of even such whit-leather beasts as mules. To save our mules, we have been obliged to make but short marches. The numerous sloughs and gullies we have met have caused us much delay. Wagons have frequently stalled on the level ground.

Our pioneers, under Lieut. Dougherty,[8] have been the hardest worked men in the command. They have not only had to march with the other men, but to work the gullies for hours at a time, mowing, cutting, and hauling wood for corduroying, shoveling earth, or carrying rocks and stones to make a road bed. These sloughs furnish practical problems in

---

[8] William E. Dougherty (1841–1915). An officer with a similar name, William W. Daugherty, was on the 1871 Eastern Survey and served during the Civil War. It is possible these two were one person.

engineering, whose solution to the officers is sometimes more vexatious than amusing. It requires no little experience, on arriving at one of these sloughs or gullies, to find the best point for a crossing. A crossing that would answer for half a dozen wagons may not answer for 250. Twenty or thirty wagons laden so heavily are enough to cut up a bad crossing so badly that every succeeding one is sure to stick. If the crossing is very bad, peremptory measures are taken at once. If wood is at hand, a sufficient number of trees are cut and hauled to the vexatious point. These logs are placed side by side parallel to the course of the gulley. If there is water enough to float them, pins are driven into the mud between them to hold them in place. A considerable quantity of brush and grass is placed on top and well weighted with dry earth. The teams and wagons then move over triumphantly. If wood is not to be had, rocks and stones which some iceberg or glacier ages ago had carted to this spot without cost find their first acquaintance with utility as the basis for a road bed. It is the indifferent places, where a half-way compromise between moisture and dryness has been effected, that are the most deceiving. They are too soft to be implicitly trusted, too hard to be thoroughly distrusted. They furnish room for experiment; and when half a dozen teams at a crossing have become hub-stalled in the mud, they furnish considerable room for regret. It is a regret which is not allied to reticence.

Bunyan's[9] Slough of Despond was a mild difficulty compared with some of those we encountered. Christian did, indeed, have a hard time of it; but he was happily on foot. If he had had a six-mule team to drive, I am afraid he would never have reached the wicket gate at all. If there be any missionary society in the East prepared to undertake the conversion of Western teamsters, one of their first efforts should be to fill up all the mudholes west of the Mississippi. There is nothing in the world that occasions so much profanity and so much cruelty to animals as one of these soft places. In crossing one of these sloughs, the teamster's cowhide is almost ubiquitous. It seems to make but little difference whether a mule is pulling or not, he is beaten all the same. He is beaten imperatively to make him go; he is beaten subjunctively [*sic*] lest he should not go. So, whether a mule is positive or negative in character is of slight account. In his own turn he gets his full allowance of lash. There are few teamsters whom I could single out as honorable exceptions to this method of crossing a ditch, but the majority of these men consider it

---

[9]*Pilgrim's Progress,* by John Bunyan (1628–88).

utterly impossible to get over such a place without vociferously damning their mules at the rate of sixty oaths a minute, as well as the ditch, the expedition, and everything else that enters into the vexation.

MILITARY METHODS APPLIED TO MUD.

When heavily loaded wagons, or as I have sometimes seen, four or five at a time, get stuck in the mud, the infantry are frequently called upon to pull them out. The men sometimes extract no little amusement on these occasions. A team of six mules and 200 men on a rope pulling a wagon make traction of no mean power. But I have seen such a team pull on a stalled wagon till the rope broke, without budging it an inch. When the rope broke and 200 men were orderly piled one on top of another like a row of fallen bricks, all idea of gloom immediately vanished. Only a well-belted man could hold his sides together. A rope breaking under these circumstances is a failure which is a sure forerunner to success. If one good laugh does not pull that wagon out of the mud, dig around the wheels a little and another laugh will.

When the first half dozen stalled teams which have discovered the treacherous road are pulled out, the pioneers go quietly to work and mend the crossing for the rest. The banks are lined with 200 or 300 soldiers watching the result. They form a jury which will soon pronounce a verdict. The pioneers work hard and well. A bridge or a filling that has to support about 2,000,000 pounds in the course of an afternoon needs to be well made. An hour or two usually suffices for a pretty bad place. The pioneers then withdraw. The teamsters on the bank mount their mules and prepare to advance. The first teamster is the object of great interest. If he is a good driver, his chances for glory are considerable. With his line in one hand and his cowhide in the other, he gives a loud shout, jerks his line, flourishes his whip, and rolls out his favorite curse. The mules understand the curse if they understand nothing else. They settle down to their work. The wagon moves. For every roll of the wheels there is a corresponding roll of shouts and curses from the lips of the voluble driver. They reach the bridge or causeway. The mules pass on in safety.

Then comes the critical moment when the whole weight of the wagon rests on the structure. But the pioneers have done their work well. The structure is firm. Yet there is a steep bank on the other side. It is a hard pull for the mules. The driver redoubles his shouts and curses. The soldiers join with "Hi, hi's." The mules are pelted with mud and stones and lashed with the driver's whip. They pull as if they were going clean out of their skin. At last the top is reached; the soldiers clap their hands. The driver pulls aside, dismounts, lets his mules blow, takes a "chaw" of tobacco, slings his whip over his shoulder, and rises several pegs in his own estimation.

But it is not the bridges and causeways but the unmade places that try the teamster's skill and discretion. It is the man who knows how to choose his place in the slough and urge his mules over it that achieves the greatest glory. I have never seen anything which quite equals the smile of self-complacency that lightens a teamster's face when he has crossed a bad place where a half dozen others were hopelessly stuck in the mud. It would take a half column to translate its significance. After such a success a quid of tobacco would never suffice in the world. If he does not take a pull of whisky at the top of the hill, he will take one before the day is over.

When we reached Heart River, near which point we expected to meet the surveyors, we found the river about four feet high and rapidly rising. Our horses forded it without difficulty, and about one hundred wagons were brought over on a bed of rocks which was made at this place last year. With the exception of several buckets which went floating down the stream, and the capsizing of one of the wagons on the bank, no loss was sustained. The water continued to rise so rapidly that we were obliged to leave over a hundred wagons on the other side, under a sufficient guard, and await the fall of the river. This took place the next morning, and all our movables were safely landed. We remained at Heart River the following day, sending Clemmo and a party out to hunt for the engineers.

A HAIL-STORM THAT KILLED ANIMALS.

The engineers had in the meantime sent a party out to find us. They reached us at night and brought intelligence of the hail-storm which had so badly treated the surveyors and their escort. Dispatches from Major Townsend, who commanded the escort, made known their helpless condition and induced Gen. Stanley to send immediate relief.

The storm occurred on Tuesday, the 24th, about 4 o'clock in the afternoon. It came on with such force and rapidity that the surveying party were taken by surprise. The party had already become well-familiar with rain and wind and thunder, but hail was an unlooked-for accession. Mules and horses can stand thunder and lightning, wind and rain, with reasonable composure. But when volleys of icy bullets come rattling thick and fast on their hide, it is not strange that their composure should forsake them and sometimes give place to the greatest terror. The only safe way is to unhitch the teams and tie them by their heads to the wagons. This storm found the surveyors and Major Townsend on the march, with a portion of the 8th Infantry and a company of cavalry. The storm is represented as most terrific. Many of the animals broke away from the wagons and stampeded. Others were overturned and tumbled on top of each other into a gulley, the animals only escaping through the breaking of the harness. The cavalry horses became unmanageable and ran away with their riders.

The whole company was completely demoralized. Major Townsend had rather a doleful experience. "My horse," he said, "became panic-stricken and utterly unmanageable and at last started off and ran away with me. How long he ran with me on his back I cannot now say; it seemed a good while; but the first thing I knew I was going over his head and plowing the ground on all-fours. Then he started off into the storm and was gone. I scrambled to my feet; but had no sooner got up than I was knocked down as if I had been struck with a club. Three times was I struck down before I could keep my feet. My hat is a sight to behold, the crown being actually torn to rags. When I got back to the command I found that many of the wagons were disabled, and one entirely wrecked. We went into camp on the spot, and sent out after the missing, and by dark, to my great satisfaction every animal belonging to the command was in camp, my horse among the first. Scarcely a man among us but received severe bruises. . . . A large yellow dog was literally beaten to death, and a number of young antelope were found dead from the same cause. When the storm ceased, the hail was from two to three inches deep on a level and a foot or more in drifts."

. . . On learning the extent of the damage Gen. Stanley ordered Gen. Custer to proceed with his cavalry to the relief of Major Townsend and the engineers, taking mechanics' rations and forage. The condition of the roads and streams was such that the main command could move but slowly. We reached Major Townsend's camp yesterday, having been detained two days by the rapid rising of [Big Muddy Creek]. . . .

THE LATEST SCIENTIFIC DISCOVERIES.

Prof. Allen, our zoologist and botanist, made a gratifying discovery yesterday. In seeking birds and eggs, he discovered the nest of a Missouri skylark (*Neocorys spraguei*), containing five fresh eggs. This bird was discovered in this region by Audubon, but there are very few specimens extant. The nest has never been found before. It is built on the ground, of dry grass, concealed from above by being arched over. It is three or four inches in diameter, about the size of our common ground sparrow's nest. The eggs are very peculiarly marked. The ground color is a purplish white, which is mottled all over by reddish specks. The larger end is quite strongly tinted with purple. Four full sets of the eggs of the lark bunting, which is very rare, have also been found. We have added 25 or 30 specimens to the Herbarium, mostly Western in prevalence. The Artemisia is found in various forms, and the common yarrow, which grows so abundantly in the East, is growing here in close neighborhood with the cactus and sage.

Since crossing the Heart River we have struck a region indicating volcanic disturbance. On a bluff near camp we found some red brick like shale with scoria and lava, and specimens showing the transition from one to the other. The bluffs were capped with quartzite, and there were indications that it extended beneath the surface. Mr. Allen traced the quartzite into the sandstone, and found in the quartzite vegetable remains in considerable abundance. . . .

Published July 25, 1873

Rosser's Diary, June 20–July 7

JUNE 19TH. At work up the Sweet Briar Creek. Stream very crooked but pretty valley and some timber. Water good and abundant. Several fresh signs of Indians but saw no Indians. Boys had a long walk to camp.

JUNE 20TH. Messenger from Bismarck got in last night. Brought an old telegram from Lizzy but nothing late. Very little timber on Sweet Briar today. Ran [survey line] 7 mile[s] today. Took a bath in this creek but the atmosphere quite cold.

JUNE 21ST. Reached the head of Sweet Briar Creek and fearing water would not be found further west we went into camp. Found coal in abundance along by the side of the line. Saw a great many antelopes but killed none.

JUNE 22ND. Rained all the A.M. and cleared off very windy about Noon and I took the cavalry and went ahead to reconnoiter the ground. Found difficult, returned to camp which had been moved 5 miles. Rained in P.M. and turned quite cold. The life now begins to drag fearfully and I want to go home.

JUNE 23RD. Not satisfied with the appearance of the line I went back to straighten it, but failed to find an improvement and returned to the [original line] and ran it out, and it proved much better as it was produced. The weather clear and warm. Feeling a little blue today.

JUNE 24TH. A most beautiful morning. Ran the line across a high ridge into another valley. Got after an elk and rode over some high hill. Found difficulty in getting a line. Train detained. Went after it. Caught in a fearful hail storm which came near destroying us all.

JUNE 25TH. So badly damaged by the storm that could not work today. Patched up wagons so as to move down into the valley. No forge and can't repair. Sent [*unclear*] to look for General Stanley and ask for assistance. Wagons broken, men bruised, and the entire company "hors de combat."

JUNE 26TH. Took party and escort in wagons and resumed work. Ran across the country over hills and valleys for ten miles. Found a practicable route but bad line which will have to be amended. [*sentence unclear.*] Very sad today, would give $500 to be at home with very dear wife and children.

JUNE 27TH. Did very little today. Sent the party out and amended the line a little, but not satisfactory and will go at it again tomorrow. The Cavalry from General Stanley arrived with appliances for repairing our wagons which we will proceed to do tomorrow. Oh, I am so terribly homesick!

JUNE 28TH. Spent the day repairing broken wagons and writing letters home thinking that I would be able to get a scout to go back to Lincoln. Visited General Custer in the P.M. Rained fearfully in the early A.M. and twice very hard in the P.M. Water very high.

SUNDAY, JUNE 29TH. Having repaired all my wagons I got an escort from Custer and arrived camp and resumed work, but General Stanley who was far behind objected to our continuing the march. I went over to see him and found him on the south side of the [Big] Muddy [Creek] with no immediate prospect of crossing for a few days.

JUNE 30TH. Produced the line about 10 miles, wrote to precious Lizzy and sent letters off to General Stanley's camp to be forwarded, but learned after sending them that he would not start the party to the rear for several days yet. Country looks a little broken ahead but I think I will be able to get a line through.

JULY 1ST. Sent party out to make some amendments to the line. Sent Eckelson and Meigs with escort 30 miles to the front to look up the country, and I finished letter to Mr. Roberts and sent him our alignment notes and [engineering] profile. Also wrote note to our precious children. I am anxious indeed to be with them.

JULY 2ND. Party returned with good news concerning line and all parties were cheered up,[10] and we moved camp about 20 miles and ran up even with camp which was 9 miles. Mosquitoes today were worse than I ever saw them and stock suffered very much.

JULY 3RD. Ran only 5 miles today on account of not knowing the country ahead. I rode ahead and examined far enough to see how to get out. Mosquitoes today are perfectly [unclear] and I had to get a brush and keep them off my horse as it did appear they would eat him up.

JULY 4TH. Had great difficulty in getting over a high ridge, the divide between the Heart River and the Big Knife. Finally got over, but on a line totally unsatisfactory. But General Custer had promised to encamp near the top of [unclear] ridge and I worked leisurely expecting to find him there but as I reached the top I received a message from him notifying me that he would not cross Heart River and I withdrew from the line and started at once to overtake the column. But to my surprise we ran our horses for two hours before overtaking the camp which had gone in the wrong direction and encamped in the wrong place. We had a hard march to get into camp and the party not pleased with the celebration of the glorious Fourth of July, 1873.

JULY 5TH. Guided the Cavalry to a good camp near Heart River and decided on the objective point on the river and then took a few troopers and went back to guide the party. Met the train, amended the line, and ran the line to about 5 miles from camp. Mosquitoes perfectly terrible.

---

[10]Rosser had found a shortcut that eliminated the most difficult and expensive section of Heart River construction. This route remains in use today and runs parallel with I-94.

JULY 6TH. Camp was not moved today and I took an escort of about one company and made further examinations and changed the line. Found an excellent line thus far and now I feel a deep anxiety in the fall of the line. I believe I will gain a good deal of distance—and the cut off line will be adopted.

JULY 7TH. Made our connection with the survey of 1871 on Heart River [at Whistler's Crossing].[11] As we are in a great hurry and as there is nothing to be gained by running a further preliminary line. Took up instruments and marched ahead with the understanding that the command would make a good march, but much to my disgust we only marched 12 miles.

Rosser to His Wife

50 Mile Creek, D.T.
June 28th, 1873

My Precious Wife:

    We have had a most tedious march indeed. It has rained almost every day and night since we started out and the ground has been so soft that it is with difficulty that wagons move. General Stanley did not leave the Missouri River with us but wasted several days, then took what he considered to be a short route, expecting to join us at a designated route, but for some unaccountable reason did not carry out his plans of March and went higher up Heart River, expecting to cross and meet us at a point further west.

    In the meantime the continued rains had raised the river and when he wanted to cross to join us he could not do so on account of the extreme high water. . . . [W]e encountered the most violent hail storm which I ever witnessed which stampeded all of our horses and mules, broke our wagons, wounded the men and placed our entire detachment "hors de combat" and we have been lying here for four days, unable to move, waiting for General Stanley to come to our relief.

    Fortunately, when the hail storm came on us we had a small detachment of cavalry with us which did not suffer very badly from the hail and

---

[11]Named after Maj. Joseph Whistler (1822–99), West Point '46; the ineffectual, hard-drinking commander of the 1871 eastern survey. The "crossing" is where the Green and Heart Rivers merge just west of present-day Gladstone, N.D.

this we sent to look for General Stanley. And last night his command came within sight of us and is now encamped about ten miles distant, but as it is raining hard this morning I don't much expect him to join us today. He sent us a few [*unclear*] and [blacksmith] forges last night and with these we are repairing our broken bones.

I am writing this with the hope I will be able to send in an escort tonight with the mail and hope to hear from you on his return. You will see a full account in the papers, I suppose, as we have a number of newspaper correspondents along with us, but to give you some idea of its intensity, I will state that every horse in the command ran away, throwing their riders except my own and [*unclear*] we managed to hold ours by getting off their backs. But my hands were cut and bleeding by the ice, my head and back was bruised, and when I stopped to change my clothing, my body looked like it had been beaten with a tack hammer. Many of the officers and men were knocked down by the ice and badly hurt, we had a large dog with us that was killed by the hail, and the prairie is covered with antelope that were killed. . . .

This storm occurred about 4 P.M. and the hail was about the size of hickory nuts and fell to the depth of 3 inches and drifted in places one and a half feet deep. Although the night was warm, the hail did not all melt until the next day. This was indeed the most terrific sight I ever witnessed. Midst the roar of the thunder, the howling of the wind, the pelting driving hail there was the wail of the men, the running of the stampeded horses and mules, and the crashing of breaking wagons. And the whole scattered over the prairie in every direction gave us a wild and tragic picture indeed. I held onto my horse for I expected when it was over if I should survive it, that I perhaps would be the only man who could go back to report the disaster.

We are all getting up again, and I have pushed my line ten miles west of this place, and if we can get our wagons repaired today we will move very rapidly in the future. And I expect to reach the Yellowstone by August 1st at the outside. . . . And as sure as God lives I will use my best endeavors to join you without delay. . . .

I am sick and tired of this work, was sick and tired of it before I began, but now am heartily so, and I will never be induced to take another such a one if I keep my present mind. The country I am running is a good deal rougher than expected, but if it does not get worse I will get a good line and thus greatly shorten the route to the Yellowstone. General Custer has marched up within a mile of us and gone into camp. . . .

55 Mile Camp
June 29th

My Precious Lizzy;

General Stanley did not reach us yesterday but General Custer did and with his command. I resumed work this A.M. and have extended my line 8 miles west of this place. I took an escort of cavalry this morning and rode over to see General Stanley who is encamped on Mud Creek about 15 miles distant. The Creek is very high and he cannot cross it, and as the troops who are with me are about out of supplies, I will have to wait in this camp until he can join me which will delay me at least 2 days....

At General Stanley's camp today I received your telegram of the 19th for which I thank you from the depths of my heart and I hope when I hear from you by the messenger who takes this, that all will be well with you. We cannot cross the Yellowstone without a steamboat, and Stanley will so inform General Terry, and I will return thence by boat, but if no boat is sent, I will endeavor to procure a sufficient escort from General Stanley, and I know I can do it, and will go to Buford where I can take a steamer to Bismarck. I am very cautious now, never go out hunting and when out of camp keep snugly by the side or in the midst of the escort.

General Custer is looking well, is very kind, and desires to be kindly remembered to you. He has no children. Several have asked me for the pictures of you [I am] carrying, but I have not parted with them yet. I have thought that I would give Custer one of them when I start home to embrace the original.

... It rained all the P.M. yesterday as I never saw a rain before in my long life, and today all small streams are rivers, and the prairie so soft that our wagons cut to the axles in moving today and my horse sunk to his fetlock at every step on a high rolling prairie where a few days ago the passing of an entire troop over it scarcely made an impression.

... Have read *Passion in Tatters*[12] and drew many parallels between the unfortunate characters and ourselves, not that our passion is in tatters by any means, but because our hearts are so often and so terribly torn by cruel separations. The vow has been registered in my heart's center *that these separations shall cease* and that the joy and happiness of constant communion shall be ours *very, very soon.*

---

[12] By Annie Thomas (1838–1918), a prolific nineteenth-century British author. She married a minister and wrote about "fallen women," the "demerits of men," marriage, divorce, and related domestic issues.

I hope you saw Mr. Roberts and his daughter Mrs. Yates. Captain Yates called on me as soon as the regiment reached my neighborhood and is a most pleasant young officer indeed. I liked the appearance of Mrs. Yates quite much as she is so much like her good old father, who is really one of the best.

June 30th [letter continues]

I have just dined with General Custer and he informs me that the party returning to Fort Lincoln leaves here tomorrow so I will finish this note and send it off. General Stanley is now 25 miles behind me and has ordered Custer to halt here until he can come up which will delay us several days. I have notified Mr. Roberts that I will reach the Yellowstone August 1st and I will allow nothing to prevent my plans being carried out unless we are visited again by rains and hail storms.

We have a large command of cavalry and it is not supposed that the Indians will show themselves until we get high up on the Yellowstone. I am painfully anxious about you and the children, and I experience the poor soldier's dream every night. "Your sweet voice on my dreaming ear melts away" at the dawning of every morn—I see our sweet little home and realize that if I have no [*unclear*] with my partners when I return I will be able to pay for it and have a deed made out in your sweet name and give it to you on condition that you bear me another sweet little child—will you accept? Bless your heart, I know you will.

Kiss the dear children and your Mama for your devoted husband.

Bradley's Diary, June 20–July 7

JUNE 20TH. Broke camp at 5 A.M. and marched at 6 with six companies of the 9th as rear guard. . . . Day cool and wet made 11 miles, camped on Mule Creek at 1:30 P.M.

JUNE 21ST. Marched at 5 A.M. in advance. Had a fine morning after a tempestuous night. As we left camp we had one of the prettiest sights I ever saw: the columns moving out together as the sun rose over the hills and the music of the 7th Cavalry band cheering up the drooping. Had some fine runs with the hounds: the greyhounds and stag hounds caught a dozen antelope. Very hot day, men and Officers suffered severely. Had two cases of sun stroke in the Battalion. Camped on Butte Creek about 4 P.M., 21 miles. Stormy and windy night: had a scare on the picket line about 10.

JUNE 22ND. Rained hard all the morning. Marched at 8 A.M. Rear guard. Camped on Butte Creek about noon: 4 miles.

JUNE 23RD. Fine cool morning. Marched at 5 A.M. Crossed a rolling country covered with fine grass: reached an unknown creek at 10 and went into camp soon after. Advance guard today: took till 5 P.M. to cross the train over the creek: 11 miles.

JUNE 24TH. Started at 5:30 A.M. Rear guard. [Wagon] train moved well in four lines until about 10 where we struck a bad slough. Took two hours to cross the train. From this point to camp found the country very bad. Got into camp about 2 P.M. on unknown creek. Distance marched 13¾ miles. Rained at intervals through the day. Very violent storm of rain and hail about dark. Upset some of the tents.

JUNE 25TH. Marched at 5 A.M. in advance. Had a tremendous storm just before daybreak. Blew down half the tents in camp. Had an easy march of 8¼ miles. Crossed Heart River at 10:30 A.M. Waited 2 hours for train to cross and went into camp. Found a very pretty valley both sides of the river, fine grass and plenty of timber.

JUNE 26TH. Lay in camp all day waiting to hear from Col. Townsend and the Engineers, supposed to be about north of us. Sent out troop of Cavalry in the morning to look for them. Reynolds [the] scout came in about dark and reported them north of Muddy, waiting for us, having encountered a tremendous hail storm which stampeded their stock, upsetting wagons and damaging stores, etc. It hailed hard enough to kill antelope and severely bruised a good many men.

JUNE 27TH. Had an early Reveille 2 A.M. in order to reach Townsend as soon as possible. Men delayed by Custer.[13] In advance today and we did not get out of camp till after 5. Crossed a rolling broken country all day, grass very fine and heavy. Camped at 4 P.M. on a branch of the

---

[13] Some of Custer's officers had hangovers and insisted on hot coffee and food. Nevertheless, Stanley waited for their stove to cool off. If he had left, Custer easily would have caught up with him. As a result Stanley missed crossing Big Muddy Creek, while the faster cavalry did. The incident left infantry and cavalry fuming at each other.

Muddy. 16 miles. Can see Custer's column across the Muddy as we go into camp from the look of the country we are off the trail.

JUNE 28TH. Marched at 5:30 A.M. Rear guard. Raining hard all the forenoon and roads very heavy. Had a hard march and made only 6 miles. Camped on Muddy River [sic] at noon. Did not get in sight of Townsend's camp, but hear Custer is there. [illegible] our half breed guide has led us off the trail and we are losing time in consequence. Violent storm at sundown.

JUNE 29TH. Lay in camp all day, waiting for the Muddy to fall. Cannot ford it anywhere. Clear and pleasant all day. Sent rations and forage over to Townsend in boats made of wagon bodies covered with canvas. River rose two feet today.

JUNE 30TH. Clear skies. Mustered the Battalion this morning. Stanley at work trying to bridge the Muddy. After a good deal of work, succeeded in making a bridge of wagon bodies filled with water kegs and covered with wagon tongues and reaches. Crossed part of the train this afternoon by hauling wagons over with ropes.

JULY 1ST. Fine day. Crossed [Big Muddy Creek] at noon and camped on north side, rain in afternoon. 1½ miles.

JULY 2ND. Marched at 6:30. Rear guard. Country broken and rolling, with fine grass. Reached Townsend's camp on branch of the Muddy at 10:30. Found them all well and glad to see us. 9 miles.

JULY 3RD. Marched at 5 A.M. in advance. Fine morning, showers through the day. Turned off from route in afternoon to find camping ground. Camped on unknown creek at 2 P.M. 18¼ miles.

JULY 4TH. Marched at 8 A.M. On left flank. Delayed near camp bridging a small creek. Comfortable day for marching. Camped on branch of Big Knife river. 14½ miles.

JULY 5TH. Marched at 6 A.M. Rear guard. Halted at 8 to bridge a creek. Delayed 3½ hours. Day fine and cool with a soft wind. Good day for marching. Overtook the Engineers at 4 P.M. escorted by the Cavalry. Reached camp on water holes at 8 P.M. 25½ miles.

JULY 6TH. Lay in camp all day resting animals and men. A pleasant quiet Sunday. Passed by the tents of the scouts in the morning and found them reading their testaments in the Dakota language. . . .

JULY 7TH. Marched at 5 A.M. In advance. Fine day, clear and cool. Crossed Heart river at 11 A.M. at Sully's crossing. Camped on the Heart at 3 P.M. 22½ miles. Reynolds and Norris hunters kill a good deal of game every day. One of the teamsters fell off his mule this afternoon and was killed by the wagon. Buried him on the bank of the river.

Bradley to His Wife

Camp on Muddy [Creek]
No. 14.[14]
July 1, 1873

My Dear Love:

 . . . We are at last across this mean little river, which has halted us for two days, being swollen by heavy rains. We bridged it with wagon bodies filled with empty barrels and poles laid across them for a roadway. We are now about 8 miles from Townsend and the Engineers and shall join them in the morning. . . .

 We have had a terribly wet time for a week and it has been the hardest kind of work to get our train along. There are indications of better weather now. You had better get the *New York Tribune* for the summer. Barrow, the correspondent here is [a] smart fellow and I think his letters will be worth reading. Save them till I get home for I shall want to see them. The *New York Herald*[15] has a correspondent here too and you may see some good letters in that. Save anything you see of interest about our expedition.

 July 2, Noon. We reached Townsend's camp an hour ago and found them all well. We all march west tomorrow morning and shall push on as fast as the Engineers can go which is not very fast at best. Rosser, the chief Engineer, says he will make ten miles a day which will be doing very well for his work. . . .

---

[14]Nos. 12 and 13 are missing.
[15]The September 24 *Bismarck Tribune* ("A Newspaper Scalawag") wrote that Harry Dashiell claimed to be the correspondent for the *Herald* and the *Minneapolis Tribune* and on "the strength of these representations . . . was allowed to accompany the expedition." Dashiell also borrowed money—at least $1,500 in today's currency—from two Bismarck merchants. One of them contacted the *Herald*, which telegraphed that they had never heard of him. But then *whom* did Bradley meet? On reaching the Musselshell, Dashiell deserted with two cavalrymen.

Camp near Big Knife River
No. 15
July 6th 1873

My Dear Love:

We are laying still this very beautiful Sunday, resting men and animals after some hard marching. . . . Our train is too heavily loaded for this country. If I had to manage such an expedition, I'd rather have less cavalry in the command for they require too much grain. The engineers are working near us, while we rest and tomorrow we move on and cross the Heart River again at what is known as Whistler's Crossing.

In about three days we shall be on the Little Missouri and there I expect to go ahead for the Yellowstone. We have had fine weather so far in July, cool and pleasant with but little rain and as we only average about twelve or fourteen miles a day, we haven't very hard work. After we strike [the] Heart River again, we shall follow it to the head and the next water will be at the Little Missouri. The country we are passing though is like [the] Laramie plains, only a little more rolling and not as high in altitude, but such a place for mosquitoes you never saw. I thought they were bad on the Missouri, but they were only visitors, here's where they live. Actually some days you can hardly tell the color of [my] horse, they are so thick on him. Sandy has been covered with lumps as large as a marble from the bites and they torment me out of my life.

July 8. We are camped on Heart River again. This morning it rained 'till 10 and after that we had it hot. We came into camp about 3 P.M. and one man dropped with sun stroke just as we got in. We did not march fast, but the sun is terribly hot when it shines out clear. . . . [T]he Engineers are doing so much better than we expected that we calculate on reaching the Yellowstone in a week from now, with the whole command. This is two or three weeks earlier than we allowed them for their work. They are a fine set of men; Rosser, Eckelson & Meigs, and they seem so anxious to get through the work as we do. Rosser says they will do their work west of the Yellowstone as fast as we can march and he says we shall be back at Fort Rice by the middle of September which will bring us home by the end of that month. This is very good and will suit us exactly. . . .

Meigs to His Family

In Camp Mud Creek
June 28, 1873

   We have only made about fifty or sixty miles of line and are delayed here by some broken wagons. Custer's cavalry joined us last night. They present a fine appearance . . . [and] look like ten thousand.

   We had a terrific storm three days since that came near costing many lives and putting a stop to the whole expedition. We were returning from work in the evening when a thunderstorm came up and Eckelson and myself who were together joined the cavalry escort. The storm became more violent still and it grew dark almost as night. The hail began to fall in larger and larger drops and began to sting considerably. The horses got very restive under their pelting and some of them unmanageable. Suddenly three vivid flashes of lightning struck the earth near us and the most terrific storm I ever witnessed fell upon us. One of the soldiers who probably had witnessed something of this kind before cried out, "A hail storm on the prairies! Go for the timber like h—-l!" and in an instant the whole troop was going downhill at a full run.

   The hail now had attained the size of large marbles and every one that struck raised a lump on the flesh. Many of us were thrown and some who succeeded in stopping their horses dismounted and tried to hold them. I had succeeded in getting off my horse and was struggling with him for he was frantic with pain when two riderless steeds rushed by and with one jerk he threw me headlong on the ground, wrenched the bridle from me and disappeared in the dense mist. The pelting of the stones was terrible and I rushed for shelter towards a little patch of brush and crawled into it [only] to find two soldiers already there. But by setting my hat high up on my head I protected it from all but the sideling stones and shielded my body by stretching my coat over my arm and holding that up high. I got some severe welts on the elbow though that left black and blue marks.

   You may judge of the force of the falling stones when I tell you that hats of felt were torn literally to rags and some stones even penetrated the brims of thick straw hats. A thick silver cup which was exposed to the storm was indented as if by falling on the floor. I think the heavy hail must have lasted ten minutes. Eckelson was exposed to it bare headed as he lost his hat and held on to his horse during most of the storm but had finally to let him go. Eck's head was completely covered with lumps as big as the end of one's finger, and his back and shoulders looked as if someone had been beating him with a tack hammer.

Our train was on the march when the storm came on and you may well imagine the result. The frantic animals rushed into deep ravines and over the rocks smashing army wagons almost to toothpicks and breaking from their harness scattered themselves over the face of the country. We managed to collect all the animals though and have patched up enough wagons to take us on....

Sioux City Journal

UP THE YELLOWSTONE

JULY 18, 1873.—The following readable notes of the trip of the supply boats up the Yellowstone are furnished us....

The *Key West, Far West*,[16] and *Peninah* were the boats [hired] for the trip up the Yellowstone,[17] and all left Fort Buford in the order named at 12 o'clock of June 25th. Each boat had on board one company of soldiers and the *Key West* had in addition one piece of artillery....

When leaving Buford orders were given for the boats to run six hundred yards apart, and as much closer as could be done with safety, but the first day out this order was violated by the *Key West* and *Far West* running out of sight of the *Peninah*. The day following the order was obeyed.

The third day out, when the boats were near Crittenden's Island,[18] where the current was very strong, the *Peninah* was swept over upon a gravel bar and was obliged to signal the other boats for assistance. The other boats returned and tied up on an island just above the *Peninah*. Sparred for some time and found it useless. In this matter the military officers gave an exhibition of red tapeism. A board of survey must be held to find how the boat came on the bar, and thus two or three hours of valuable time was lost.

---

[16] The *Key West* (built in 1860) and *Far West* (1870) were captained at different times by the LaBarge brothers and Grant Marsh. The *Far West* was 190 feet long and 33 feet wide with a freight capacity of some 250 tons; the *Key West* was far smaller at 169 tons.

[17] The *Josephine* also brought up supplies. Named by Grant Marsh for Stanley's oldest daughter, the *Josephine* was built in the winter of 1872–73 and designed for upper Missouri/Yellowstone use. With a 300-ton capacity, it was 178 feet long, 31 feet wide, and had a draft of only 1 foot. All told, the four made nine trips to Glendive Creek.

[18] Halfway between present-day Sydney and Glendive, Mont.

The river was rapidly falling and every moment's delay made the matter worse. It was finally decided to take all her freight off and make another effort to release the boat. The military detailed a squad of soldiers to assist the crew in lightening and the crews from the other boats also assisted. The freight was all taken off and [rowed] to an adjacent island or sand bar, and by 10 o'clock in the evening the boat was again afloat. We then tied up to the bank for the night and at daylight next morning we dropped down to the freight pile and commenced reloading, and at noon all were ready to move up the river again. The boats ran but a short distance and all tied up for the night.

Next morning all started together and ran along nicely until 4 o'clock when they came to Jewell's Chute where they were obliged to . . . pull over. The *Key West* passed over safely, but when the *Peninah* was about half over, the [steam capstan][19] gave out and she dropped across the chute breaking her capstan, wheel, and flanges, and she was again obliged to signal for assistance. The *Key West* tied up and sent a yawl down to ascertain what damage had been done. Those who came down returned and reported to the commanding officer, and by him, it was ordered that the *Key West* and *Far West* should proceed to Glendive, return and take the *Peninah's* freight, if in the meantime she should not be able to proceed up the river.

On the morning of the 29th the two former started up the river, and on the 30th the *Peninah* was again ready to start. She ran but a short distance and tied up to chop wood.

On the night of the 30th one of the pickets saw, or imagined he saw, two Indians creeping along the timber. He fired an alarm which was answered by the sentinel on the hurricane roof of the *Peninah* firing two shots. The long roll was beaten and all got up in a hurry. The soldiers got on the bank and formed in a line and one of the officers went out to the picket line to ascertain the cause and found as above stated.

During the excitement some ludicrous scenes occurred on board the boat. The first engineer, a Hoosier, from Hoop Pole Country—himself a fair specimen of hoop poles—rushed out of his room, trousers in hand, yelling, "Indians! Indians!" while from the next room came a female fifteenth amendment, who had on no more wearing apparel than the law allowed, and appealed to a certain officer on the boat not to leave her *now*.

Nothing further occurred on the trip. . . .

---

[19] A winding rope "employed in hauling riverboats over bars or snags." *Shorter Oxford English Dictionary*, 5th ed., 2002.

Konopicky to His Parents

Est. 60 miles from Ft. Rice
June 21, 1873

At 5 o'clock [A.M.] on the 20th, the large military column left Fort Rice for the West. The order of march was as follows: in front, a unit of infantry; then the Commander and his officers with the five of us in the Scientific Corps, like princes on horses; then in two rows the innumerable wagons with a canon in front of each row; to the left and the right of this column about 300 steps away rode the cavalry in long lines; at the end of the caravan was a large herd of cattle intended for food. Far ahead rode some 40 Indians scouts in the form of a cross. In this order we rode for 5 hours until halted by a thunderstorm. We pitched our first camp in a pretty valley surrounded by hills. It was high time as far as I was concerned, because I was so tired [from] riding that I could hardly stand. The following day I forgot about riding and sat in our wagon.

Our second night we camped on the prairie. The night we spent there was memorable. There was a thunder storm, such as I have never imagined, from 9 in the evening until 5 in the morning. The sky was a sea of fire with lightning bolt after lightning bolt, thunder after thunder. I heard the lightning striking to the left and right. It was as bright in my tent as if I had a gaslight lit. Every moment I thought my tent would catch fire or the wind and rain would come streaming in. Despite all of that, the trumpets sounded reveille at 3 in the morning (as every day) and we continued in the lightning and thunder (naturally I am in the wagon).

During the trip countless bounding antelopes were shot. The soldiers had received permission from General [Stanley] to relieve the tedium of the march by shooting. Unfortunately the soldiers were not very good shots, otherwise they could not have succeeded on the second march in killing 2 mules and a dog, one of the 20 that accompanied the caravan, instead of antelopes. Just now another big thunderstorm started, so I will stop writing.

It was a sea of fire like the last time, but with the difference being that this time water came streaming into my tent. At least it didn't last as long. In the morning, a windstorm blew down a few tents. Today, we reached a valley surrounded by sand dunes, a quick stream surrounded by trees (Heart River). We had to ford with all of the horses and wagons and pitch camp on the other bank. As before, fording the river with some 250 wagons took a long time.

The heat here is terrible. During the second march, 8 men collapsed and had to be carried on by wagon. Thank God, I am in good health so far, and I found the baths I enjoyed in the Heart River very refreshing. I can also say that I am very fortunate to be on such a rare journey. I feel lucky to have been given the opportunity to admire God's nature from one of the most untouched parts. Actually, when I sit on a mountain or a hill making sketches of the landscape, I feel I am hearing the cavalry band playing something from Beethoven or Schubert, etc., or whatever the case was, I hear the Blue Danube Waltz.

Camp life is tedious. Our tents are always pitched near the General's, who gives us a great deal of attention, and a Major named Lazelle, a very likable American, is always by our side. The staff and senior officer visit my tent daily to see my sketches.

We shall continue early tomorrow morning and I will have my horse saddled. I am told that it is about another 4 weeks to reach the Yellowstone River. Then we will follow the River further. The postal service has become pretty bad. It is now in the form of some Indians, who have taken over the task of taking letters through the wilderness. . . .

Our daily marches since the 24th have been fairly tiring, that is, for the horses and mules which find the ground, soaked by continuing thundershowers, heavy going. The ground is a soil in which one doesn't find sand. On top of that are the swampy waters, which we must cross, where the mules often sink in up to their bellies.

For the last two days, we have camped on a rise from which there is a marvelous panorama of the many barren hills. Actually, it is a desolate picture, much like a desert. I could have been easily able to count the trees we have seen since Fort Lincoln. The reason for our longer stop is that we have to build a bridge to cross a small stream ahead of us. Today, the 30th, we are momentarily expecting the order to break camp, but I don't think we will start today. I don't at all mind such a longer stop, because I can catch up on my sleep. However, that probably means that we'll have to be up at 1:30 A.M. like three days ago.

In the evenings when I return to camp with my landscape sketches, I always have officers visiting me to see the sketches (quite honestly, these visits don't make me especially happy because the sketches get smeared from being passed hand-to-hand). In the evening the day before yesterday I had a visit from our likable Major Lazelle (I say "our" because he is quite at the disposal of the Scientific Corps, in which he takes a great interest). That evening he was accompanied by *Officer* Grant (son of the

President, who is also with the expedition) and several other officers who looked at my sketches.

90 Miles from Fort Rice
June 30

...What a contrast between Vienna and the corner of the earth where I find myself [today]. There is a collection of people from all over the world, here in a desolate land empty of living things except antelopes and a few birds. Only 100 miles further on and we will have the pleasure to see Indians. To be sure, camp life is not exactly restful and especially the 1,500 mules with their—?—like braying bring much life to the area. I spend a lot of time especially with Pywell, the photographer; he amuses me with his jokes. The continuing thunderstorms are unpleasant; I had no idea of their strength. So far we have always come through them all right, unlike the detachment of engineers traveling 25 miles ahead of the column. Four days ago a hailstorm with ice pieces the size of chicken eggs half destroyed their wagons, so the General had to send workers[20] with a detachment of cavalry to repair them. It is very unusual weather here. Last year, for example, there was almost no rain.

---

[20]Blacksmiths, wheelwrights, and the proper equipment.

CHAPTER 5

Mid-July

By leaving Fort Rice in mid-June, not later in the summer as had the previous surveys, the 1873 expedition soon found itself floundering on the muddy Dakota prairie. Not until the rains slackened and Stanley left the valleys of the Heart River and its feeder streams did the pace improve. As the expedition reached the drier plateau, rapid progress followed, and Custer and Rosser arrived on the Yellowstone July 15. It was during this period that the Stanley-Custer relationship reached its lowest level, with Stanley placing Custer "in arrest" for a day.

Cynics believe snafus (situation normal, all "fouled" up) to be routine in army life, and 1873 proved no exception. An integral aspect of supplying Stanley was sending ten shiploads of supplies up the Missouri to Glendive Creek. Here, within a three-sided triangle, "Camp Canby"[1] was established. Canby faced the Yellowstone to the west, Glendive Creek and a forty-foot-high steep bluff to the south, and a marshy, flood-prone area to the east.

However, as Custer and Rosser immediately discovered when first seeing Canby, there was one slight problem: wagons couldn't reach it. This oversight meant finding a new camp—"Stanley's Stockade," ten miles upriver. Thus, day after day, every pound of supplies had to be reloaded, sent upriver, unloaded, and the process repeated.

Chapter 5 begins with Barrows's description of crossing Big Muddy Creek and the beauty of the northern Dakota badlands. Rosser's invaluable diary cites his acting the peacemaker in the Stanley-Custer dispute, Stanley's horrendous drinking, and Custer's desire to arrest the drunken commander—one of the expedition's fascinating what-ifs. Also included

---

[1] Named after Gen. Edward Canby (1817–73), who had been murdered during peace negotiations with Modoc Indians that April.

are Bradley's diaries (I have remarked earlier about the coincidental disappearance of his letters from the period of Stanley's drinking), a Konopicky letter, and two newspaper descriptions of Custer's arrest.

One of the survey's interesting personalities was the president of the Winoma, Minnesota, normal school and a protégé of William H. Seward, William F. Phelps (1822–1907). A leader in educational reform, Phelps was an early president of the National Education Association and a founder of the normal school movement (which evolved into teachers' colleges). But why he joined the Yellowstone survey is unclear.

Rosser nicknamed Phelps "Fagin," and one surveyor recalled him as "a tall, gaunt man with long legs, deep set eyes, with little tufts of [a] bristling beard sticking out of his face. . . . [His] ears were large and set out from his head. He rode a small pony and [his] long legs dangled almost to the ground. He slept in our tent, never took off his shoes, and snored like a Turk."[2] Phelps's insightful description of Custer and Rosser was to catch the public's fancy and be printed around the country.

---

[2]A. L. Berry, "The Yellowstone Expedition of 1873," Rosser Collection, University of Virginia, Charlottesville, #1171-g-h-j, box 12.

Barrows to the Tribune

YELLOWSTONE EXPEDITION

TRAVEL IN A NEW AND WILD REGION

A SWOLLEN RIVER BECOMES A SERIOUS OBSTACLE— NO PONTOONS PROVIDED—NECESSITY THE MOTHER OF INVENTION—A NEW USE FOR WHISKY KEGS—THE SPLENDORS OF WESTERN SKIES— THE BAD LANDS OF DAKOTA.

HEART RIVER, DAKOTA, JULY 8.—Diametrically, there is a stream in Dakota which measures about thirty feet. Chronometrically, it is just about thirty miles. That is, if you take a tape-line and measure from bank to bank at the top it is just about thirty feet. But if you take its width and depth, and multiply it by ten days of heavy showers, and multiply the product by 250 heavily laden wagons without pontoons or timber, you will get the time value of crossing it. In our case it was over two days. In that two days we could have made thirty miles. This stream or miniature river is known as the "Big Muddy." The adjective is the only necessary part in the description of a stream in this part of the country. The noun is the generic or family name. At least it ought to be. The streams are of the earth earthy. Temper a cup of strong coffee with a little milk and you have the prevailing color. Gen. Custer, whom Gen Stanley had sent with the 7th Cavalry and a few light loaded wagons to afford relief to the hail-pelted engineers, had arrived at this stream the day before. He had found the water too high to cross, and built a bridge in the afternoon on which he had crossed his troops and wagons. Through a mistake of our guide and the wretched condition of the roads our heavy train did not arrive at this stream until the following day. Our mules were almost exhausted by their pulling. It was deemed unadvisable to attempt crossing that night. It was doubtful whether we could cross without building another bridge. The stream that night decided the question for us. The heavy rains swelled it considerable. Its level rose high and higher. Somehow it seemed to take a fancy to Gen. Custer's bridge.[3] The bridge seemed to reciprocate it. There was a collusion and finally an elopement. I saw the water creep up closer and closer, and at last lift the bridge from its feet and carry it off on its bosom. We had no

---

[3] About 2½ miles south of present-day Almont, N.D.

objection to the amours of this bridge; but it was unkind of it to desert us just as we needed it most. The stream had risen so high that it was impossible to bridge it again on account of its increased width. We had no timber long enough.

It is a fact known, I suppose, to the United States Quartermaster's Department that an army wagon laden with 5,000 pounds will not float in ten feet of water. It is unreasonable to expect six mules to swim with such a weight. It was with reference to this want of levity in loaded army wagons that pontoons were invented. But, notwithstanding this fact, the largest expedition since the war started off on a journey of several hundred miles across a new country without a single pontoon. The reason is that the country has a dry reputation. This year, however, it has belied its name. A few pontoons would have neutralized the falsehood. They were asked for, but there was a knot in the red tape somewhere, and we did not get them. With pontoons, our whole command might have crossed in three or four hours.

AN EXIGENCY THAT DEVELOPS INGENUITY.

Gen. Stanley is not a Micawber.[4] He never waits for something to turn up if there is any chance of turning it up himself. Besides the necessity did not admit of much delay. Major Townsend's camp had been moved ahead seven or eight miles. They needed forage and rations. The cavalry in advance needed rations also. If there is any place where ready practical talent finds ever fresh scope for exercise it is in active army life on the Plains, especially in connection with such an expedition as this. . . . But if necessity compels, [the soldier] can get more comfort and more work out of a pound of beef than any one outside of his profession—almost like the protozoans, he can extemporize a stomach to meet his needs. If he has a full ration he is sure to be equal to it; if but a half a one, he can make up the rest by growling. His domestic economy is reduced to first principles. A spade will furnish him a cooking stove. A tin can will make a candlestick. A bayonet thrust in the ground will serve for the same purpose, or answer to pin down his tent. I have seen leavened bread, as fine as any you can get at a New York bakery, baked in an oven dug in the clay on a hillside. If it be true, as Socrates said, that they who have fewest wants live nearest to the gods, then the soldier comes

---

[4]Wilkins Micawber, the perennially optimistic job seeker in Charles Dickens's *David Copperfield*.

materially near to a godly life. I am afraid that it is only in this sense that the claim can be maintained.

How should we get over the Big Muddy? It was a problem for an engineer. The problem was there, but not the engineer—I mean a member of the engineer corps. I find, however, that army officers as a general thing do not place a very high estimate on our engineer corps. "Why did not a regular army engineer accompany this expedition?" I asked of an officer. "Oh, he'd be afraid of getting sunburnt. Besides, we can get along better without them. They can't work unless they have everything just so. They are good to stay in the office and make maps, and that is about all they are good for." Be this true or not, we happily have two men with us who are better than a dozen desk engineers. I refer to Gen. Stanley and Lieut. Ray, our Chief Commissary [officer]. Gen. Stanley is a thoroughly educated officer, and has had a wide experience on the Plains. He has a natural talent for his profession, and unites excellent powers of observation with rare judgment and ability to command men. Lieut. Ray is eminently a practical man. He has seen life on the Plains in all its aspects, and served with great credit during the war. He has passed through many trying experiences, and never but once was caught in a place that he could not get out. This happened near New York, and brought him worthily into public notice.[5] While stationed at Davis Island, he went out in a boat one stormy wintry night to rescue a party who had been caught in the ice. He reached them, but was unable to return to the fort. The floating ice carried him far out into [Long Island] Sound. It was not until the next day that they were taken off, with frost-bitten hands and feet. . . .

In fact we were all in a state of extreme cogitation. It was one of those places where a good deal of thinking may be done to very little purpose. The first thing to do was to get forage and commissary stores over for Major Townsend. This difficulty in a stream twenty-five or thirty feet wide did not present the magnitude that the transit of our wagon train did. If we could get the forage on the other side, Major Townsend could send back some wagons for it.

A SERVICEABLE BOAT FROM SLENDER MATERIALS.

There were two propositions. One was to construct a rope-line elevated railway, with an aerial car to be pulled back and forth by hand. Another

---

[5]The winter of 1869–70. Davis Island was a seventy-eight-acre military post guarding New York City, the land adjacent to the city of New Rochelle.

proposition, made by Lieut. Ray, was to extemporize a boat. All that he asked was a wagon bed, some rope, and a tarpaulin. The railroad finally gave way to the boat line.... [However,] there was much distrust in the boat. Prophetically it sprung a leak several times before it was launched. Hypothetically, it was so laden with contingencies that there was no need of launching it at all. But Lieut. Ray said, "I shall show my faith by my works." Gen. Stanley answered "Yea and Amen." A wagon body was dismounted. It was wrapped on its bottom and sides in a heavy "paulin" which was firmly secured by ropes. It was the work of a few minutes to make it and launch it. The heavy canvas effectually kept out water. It was safely navigated to the other side. Gen. Stanley and Lieut. Ray were among the first to cross the rapid stream. A number of men, stationed on each side, easily pulled the boat across, receiving no little aid from the current. On this little craft we could safely put 1,000 pounds of forage. It solved one element in the problem. In five or six hours we had ferried enough stores and forage to supply the . . . surveyor's escort.

The next question was, How should we get over our heavy train and teams? On Sunday afternoon, while we were crossing our forage, the stream showed a decided tendency to fall. Its level was quoted at various stages during the day, but toward night our stake showed a decrease of three or four inches. It is characteristic of these streams that they rise and fall very rapidly. If the rise is rapid, the fall that succeeds is most likely to be so too. As the stream was decreasing, it was hoped that by morning it would be sufficiently low and narrow to permit building with the short timber that grew along its banks. The hope was delusive, the expectation vain. Instead of going down, the stream, refreshed by some remote tributary, went up to its original height. When Monday morning came, we were just as far from the other side as the day before.

A BRIDGE OF WHISKY KEGS.

Gen. Stanley was determined to wait no longer. The Commissary, Lieut. Ray, once more solved the problem. He offered to build a bridge and cross the command. But how could he build a bridge without timber, pontoons, or lumber? We have in this expedition over 100 kegs. When they contained whisky the bung always leaked. They finally leaked dry and became hopefully converted. These temperance kegs have been heretofore carried on the wagons. Ray now proposed to carry the wagons on the kegs . . .

Ninety-six empty kegs were accordingly ordered to report at the stream. Four wagon-beds were dismembered of their covers and wheels. Twenty-five or thirty men were then set to work to bung and plug the kegs. Only those thoroughly coopered were accepted; the dry and unserviceable ones were rejected and their places supplied by others. Each of the wagon-beds just held twenty-four kegs placed on their ends side by side in three rows. Some timber was, meanwhile, cut into poles, and one pole placed lengthwise over each row of kegs to keep them in place. Ropes and chains were passed over the poles and completely around the wagon-beds, so that the booms and kegs were firmly secured. The wagon-bodies were then dragged down the bank and launched in the water bottom-side up—that is, with the kegs down.

Each wagon-bed was then floating upside down, upheld by twenty-four air-tight kegs. These extemporized floats were then moored lengthwise in the stream. The next trouble was to lash them securely side by side. This was no easy matter; but Mr. Ray . . . had one man on the first float with him, and with his coat off and sleeves rolled up, lashed them together with his own hands. The shores were lined with officers and men watching the experiment. Very little confidence was expressed in the structure by the wagon-masters and teamsters. "I tell yer," said one, "this here thing aint goin' to work no how. Jest as soon as one of them big wagins gits on that bridge the old thing'll sink an' wagins and mules will go too." The destination indicated was much further than the Yellowstone. Some of the less experienced officers considered the whole thing as a foregone failure. The bridge, like the boat, was born under a cloud of doubt.

By means of some wagon-reaches and plenty of picket rope, the floats were finally fastened together. If we had had some plank now to place across the inverted wagon-beds, one bridge would have been complete. If we had only brought one plank to each wagon in our train we should have had ten times as much as we could use. But we had not a single available plank or board. We were compelled to fall back on our extra wagon-tongues and reaches. These were placed side by side across the wagon beds; they did not make a very even floor, but it was the best we could do. The bank had been previously cut away to form a gradual descent. It was easy to connect the float with the shore.

The bridge was done. I imagine that Lieut. Ray watched with considerable interest the crossing of the first wagon. To secure greater safety the mules were detached and led down the bank and over the bridge.

The wagon-wheels were locked and the wagon gently let down to the bridge. A detail of men on the other bank then took the rope and pulled the heavy-laden vehicle across the float and up the opposite bank. Under the great weight the bridge sunk to the water's edge but no further. The wagon crossed in safety. The success was repeated 250 times. When finally all our teams and men were safely landed on the other side without a single accident, without losing a pound of forage or a single piece of hard-tack, the bridge had published its own triumph and the doubters were silent.

Such a bridge was not a new undertaking or achievement for Gen. Stanley. In his 20 years of army life, all of which has been spent in active service, he has become familiar with every known device for managing a train. I doubt if there is an officer in the service better qualified for the difficult work of conducting a train through almost impassable places. There are few who have such resources of judgment and experience. On this trip he is fortunate in being seconded by a competent staff. The mechanical success of the work was due to the perseverance and ingenuity of his Chief Commissary, who constructed it mainly with his own hands. Our little bridge will never attain the fame of the Victoria;[6] but it shows what may be accomplished by putting this and that together, brain-wise and otherwise.

BUTTES AND VIEWS FROM THEM.

JULY 13.—Having crossed the Big Muddy we pushed on to Major Townsend's camp, and arrived there on the 2d instant. Major Townsend, as previously noticed, had been detailed to escort the engineers from Fort Lincoln until the main command should overtake them from Fort Rice. Major Townsend had gone in camp and awaited our arrival, after suffering from the hail storm, and the cavalry had gone on with the engineers. Spending one night at the camp we pushed on the next day with our train to overtake the latter. It was a morning of rare beauty. The day was born without a blemish and ushered in by the sweet carol of birds. It is one of the mitigations of breaking sleep at 3 o'clock in the morning that we greet the day on the very threshold of dawn. No account of the scenery of the plains can ignore the scenery of the sky. The sky is always half of the picture. There is the same intimacy here that exists

---

[6]Barrows was referring to the bridge in Wales.

between the sky and sea. It is this companionship which makes the clouds seem neighbors to the soil. Finer cloud pictures I have never see. Royal banks of white fleecy snow seemed piled up in measureless heights against the arch of the sky. On the cool breeze we could almost catch their breath. Silent, restful, baby clouds slept in cradles of calm over our heads. Now and then great fleets of white, vail-like mist moved along like an armada, ultimately destined to form "in close order," descend and dispute our comfort. They would tell the streams of our approach. We should bridge these clouds on our way.

One vision of clouds and color I shall never forget. A gentle shower broke upon us one afternoon. It was soon over, and the sun burst through the clouds. As if by magic a magnificent rainbow sprung with a radiant arch across the western sky. Prodigal in color and perfect throughout, it was a sight of indescribable beauty. But within this gorgeous frame was the picture. From one foot of the arch a long black peninsular of cloud faded away to the center. A high cloud bluff of purest white crowned with a resplendent gold arose from this peninsular of shade. Together the two formed a beautiful perspective framed with the spectrum. Riding along at the head of the column toward this scene it seemed as if the curtains of heaven had been drawn and the New Jerusalem unveiled. For a moment we were only pilgrims journeying to the Heavenly City.

This morning our cavalry band had gone on in advance with their regiment, but we marched to a sweeter and fresher music. All the larks and buntings of the plain had joined in an overture. It was a bird jubilee. No orchestra ever played sweeter. Our trail led us over a level section of the plains less rolling in character than that which had preceded it; but still varied in different directions by peaked buttes and insular hills. One of these buttes to the north was especially striking from its bold and lofty character. A few of us ascended it, and obtained an extensive and beautiful view. A grassy plain stretched out for two or three miles, with only here and there a gentle swell to break the surface level. For miles to the east this gentle undulation was repeated. Looking over this plain to the south, the eye was arrested by two tall sharp buttes of equal height, standing a short distance apart on a line east and west at right angles to the observer. They had evidently been connected at one time, but some great water scoop has hollowed a pretty vale between them. From the point on which we stood, these twin buttes formed the door-posts to a delightful vista, extending clear to the horizon.

The country to the north was broken up into a thousand conical hills, ranging, apparently, from 40 to 75 feet in height. Most of them were sharply pointed and steep-sided. Water and wood had rounded them into symmetrical forms. In the distance they seemed like so many sugar-loaves dropped on the plain. The butte on which we stood was well-clothed with grass, with here and there a sprinkling of flowers. Tooth-shaped outcrops of sandstone firmly set in the jaw of the rock appeared at intervals on the surface. In two or three places the strata were lifted, folded and much displaced, but the general stratification through this and most of the other related buttes we found to be regular. So far we have had little to do with stratigraphical geology. The exposures have been mainly regular sandstones or clays but slightly indented and with little variation in disposition. Except the gastropods noted in a former letter we have found no fossils worthy of mention.

RED AND WHITE MEN'S CAMPS.

As we descended on to the plain, the stakes in the ground showed that we were on the trail of the engineers and in the path of the future railroad. We could hardly help contrasting the Yellowstone train of to-day with the Yellowstone train of the future. It takes 1,500 mules to draw our train. Three energetic locomotives could do it with less distress. In ten hours the coming locomotive will travel as far as we shall have journeyed in 25 days. In all our trip of 1,000 or 1,200 miles we shall not see a single house. The railroad will bring houses and inhabitants. This survey is the harbinger of civilization. It is the John the Baptist that makes the way straight. It will be an easy thing to "do the plains" then with sleeping-car and dining-room comforts.

We soon struck the site of a former camping-ground. There were indications that it was not an Indian camp. The debris of an Indian camp is very different from that of a white man's. The Indian builds his fire-place of stones; the soldier, to prevent the grass from taking fire, digs a square fire-place in the ground about six or eight inches deep. The Sioux teepee is quite different from the soldier's tent, and by the marks they leave in the ground one can readily be distinguished from the other. The teepee is a lodge composed of several poles placed in the ground in a circle and bent to the center at the top; skins, robes, and canvas, are wrapped around them to afford shelter. The fires are built in the center of these lodges, the smoke issuing from an opening in the top. Bones and sinews, scattered on the ground, show something of the bill of fare. When on a journey the Indian usually carries his lodge poles with him. They are tied on each side of his horse like a pair of thills, one end dragging on the ground. They are called tracers. On these he packs his provisions, equipments, and spoils. The trailing poles leave a mark on the ground easily recognized.

At this camp the signs of tracers or teepee were wanting. There were heavy wagon trails and marks of "A" and shelter tents.[7] Here and there posts were left standing in the ground. Ashes and burned wood suggested an extinct kitchen. The prairie carpet was disfigured by tent-holes, trenches, and fire-places. From the envelopes and letters scattered around, one might have inferred a deceased post-office. A few empty bottles were left lying on the ground. They were not medicine bottles. They were very dry. Their history was evident. They had become exhausted on the march, and had to be dropped. Near the bottles there were carte di vistae [*sic*][8] of two royal personages. They had evidently slipped out of some teamster's "album." One was a portrait of the King of Spades; the other was a picture of his Queen. Some tin fruit cans found around the camp showed that the Indians had not been there since the whites. Indians have a passion for these cans, and are sure to pick them up when they get a chance. If the white man leaves food behind at a camp, the Sioux seldom take it. Their timidity in regard to food may have been occasioned by an unfortunate meal which one of the ten tribes made a few years ago. . . .[9]

Near this camping-ground, which had been occupied by our advance party escorting the engineers, was a curious rocky ledge running from north to south, some 50 yards across the rocky plain. It was a reddish sandstone, broken up here and there into boulders whose corners had been worn and rounded by the action of water. The wall of rock, from three to five feet high, had escaped leveling by the denuding floods. Its appearance suggested the man who began to build, but had neglected to count the cost thereof. We pursued our march until 4 or 5 o'clock in the afternoon, and then camped on the banks of a stream, which required bridging.

---

[7]A-tents were used by officers. The meaning of "shelter tents" is perhaps a "pup" tent or simply a poncho(s) hung between poles.
[8]A *carte de visite* was a business card–sized photograph.
[9]Here Barrows repeats, as true and taking place just a decade earlier, a mid-1700s story of trappers placing arsenic in sugar and making sure that Indians "stole" it. But, of course, it also could have been done multiple times.

Our Fourth of July was not distinguished by any special observance, except a noon salute of 13 guns and a march of 18 miles. It was, perhaps, a more patriotic way of observing it than if we had idly stayed in camp all day and burned more powder. I think, too, that a half a day in camp was as much as was expedient for purposes of private observance.

On the 5th of July we came to a tributary of the Big Knife River which cost us some three and a half hours work in bridging and crossing. In the afternoon we descended from a high divide into a beautiful valley abounding in grass and wild rye. We could see the engineer's flags about a mile off. By 2 o'clock we came up with them and found Gen. Rosser, the chief, in the field. The cavalry had gone into camp near the Heart River, some six or seven miles distant. There was no wood or water this side of their camp. We were obliged to march on. It was 7 in the evening when we made the camp. The bugle had never sounded more musical than when it sounded the halt that evening. We had marched 22 miles that day though much interrupted by the streams and sloughs. We were yet three or four miles from the Heart River. Water was obtainable, however, in a small creek near by, and our Indian scouts soon brought in some wood. It was nearly 10 o'clock at night before our dinner was ready, but it was 18 hours since breakfast with only a light intervention in the shape of a lunch, and somehow the antelope which "Given" our dog had caught seemed the finest broil we had tasted on the trip. . . .

THE SURVEY OF 1871 TO BE UTILIZED.

Our meeting with Gen. Rosser caused a decided change in our plans. . . . It was supposed that we could not make more than 4 or 5 miles a day at most, on account of the survey. It was calculated that it would take us 50 or 60 days to reach the Yellowstone. Supplies were taken for 40 days, and when 50 miles from Fort Rice a wagon train was sent back for additional supplies, which it was supposed we should need before we reached the Yellowstone. When we reached Gen. Rosser we learned that the survey had met with unexpected success, that instead of taking us 30 or 35 days more to the Yellowstone, we could make it in ten days. The news gave us no little surprise and pleasure.[10]

As mentioned in a previous letter, while this is the first survey of the country from the Yellowstone River to the West, it is the third survey

---

[10] Stanley believed Rosser had purposely withheld the information. While Rosser's timing and tact were terrible, Stanley's angry reaction bordered on the paranoid.

from the Missouri River to the Yellowstone. The survey of 1871 followed the valleys of the Heart River and Glendive Creek. It was finally adopted as the general route. It ran a little too far to the north to be acceptable, and the survey of last year was undertaken to cut off this angle, making the mouth of the Powder River the objective point. This route was found impracticable, and this year another survey was undertaken. Gen. Rosser, after surveying the country from the Missouri to the Heart River, has found a way of connecting the survey of the year with the survey of 1871, rejecting last year's survey entirely. By the survey of 1871 the Heart River had to be bridged 100 times, averaging 150 feet each time. In this country, where there is neither timber nor building stone, this was a formidable objection. Stone and timber would have to be forwarded from Minnesota These bridges alone would have cost $300,000. According to the present route there are now only 11 crossings, and not over 50 feet of water-way. Under this determination of Gen. Rosser, we could push through to the Yellowstone River as fast as possible, the engineers having nothing to do until we reached the "Bad Lands," when, with a small detachment, Gen. Rosser could straighten the road at that point.

It would have saved considerable expense and no little trouble had we known this at the start. Thirty days' forage and commissary stores would have been sufficient, and the extra wagon train sent back to Lincoln would have been entirely unnecessary. A smaller number of wagons would have sufficed and our mules would have had an easier time. The Yellowstone is now estimated to be about 150 miles distant. We hope to reach it before the 20th inst. We are anticipating, however, some trouble on the "Bad Lands," which lie between us and the Little Missouri.

CHARACTERISTICS OF THE ROCKS.

"BAD LANDS," Dakota, July 17.—We pushed on toward the Little Missouri [and] found the water low at the head of the river. On the bluffs and buttes nearby there was an abundant exposure of rock, showing the same friable sandstone which we had met day after day since leaving Fort Rice, horizontally stratified, and containing no fossil remains. Large masses of the rock had broken off and rolled down the steep sides of the buttes, and lay piled in confused forms. The stone is unfortunately too soft for building purposes. Here we found larger varieties of spiderwort than any we had seen, bell flower grew in abundance, and the wild flax seemed more lusty. The yucca which we had met several days before grew sparsely on the hills. It is a woody-stemmed plant of the lily family, growing two feet high, with a whorl of thick, prickly, palm-like leaves, from the balls. The root is very white and stocky, and furnishes a good lather which is used by the Indians and Mexicans in lieu of soap. Above all the other plants in profusion and beauty were the wild roses.

For some days of our march we traveled literally through a bed of roses, representing all shades of pink and red and delightfully fragrant. One creek in the line of the survey was so full of sweetbrier that the engineers named it Sweet Brier Creek. Buffalo grass, wild rye, wild oats, and wheat we found in various alternations with other grasses. Prof. Allen was fortunate in securing six Missouri skylarks, a rare bird previously noticed, whose nest was obtained for the first time on this expedition. Other species of birds common to both the Eastern and Western States were added to the collection, and some new varieties of plants.

All along our route we have met with petrified wood in considerable quantities. It abounds along the Missouri, and is readily found for 200 miles west. Some of the specimens were very beautifully silicified. A fine collection of fossil woods could be made from this section of the country alone. The specimens present considerable variety in size and appearance. In some places we found stems of trees several feet in diameter two or three feet above the ground, perfectly petrified but clearly showing the grain of wood. Lignite[11] in considerable quantity is found along the Heart River, and has been successfully used by the engineers for fuel. Gypsum and silinite appeared in small quantities.

On the 8th inst. We came to a singular formation, unlike anything we had passed. A large bluff of white, argillaceous sand, perhaps 60 feet in height, and from 75 to 100 yards in its longest diameter, occurred upon the plain, directly in line of our trail. Entirely denuded of vegetation, it stood out like a bald, naked island of dazzling white, fiercely reflecting the hot rays of the sun. In shape and appearance I can compare it to nothing but a huge pile of mortar, a bleak, unproductive union of clay and sand, unfit for man, beast, or vegetable. Its top was covered here and there with small, dark, conical sand hills, along its base were traces of iron.

As under our more rapid rate of marching it would not be possible for the supply train that returned to Lincoln to overtake us before reaching the Yellowstone, Gen. Stanley deemed it advisable to leave behind 100

---

[11]Brown coal.

men under the command of Capt. [Edward] Pearson of the 17th Infantry, to strengthen the Quartermaster's escort. It was feared that the Indians, if they attacked us at all, would trouble us in the Bad Lands. Under this disposition of the troops both trains would be adequately protected, Col. Baker having, in addition to the 100 infantry, two companies of the cavalry under the command of Capt. [G. W.] Yates. On the 8th of July, when about 164 miles from Fort Rice, breastworks were erected and Capt. Pearson was left with orders to await the arrival of Col. Baker.

The following day Gen. Rosser, with his assistants, under escort of a squadron of cavalry under command of Capt. Bestine,[12] left the main command to correct his survey of the "Bad Lands" and to rejoin the main command at the Little Missouri River. The weather was varied that morning by a fog so heavy that it seemed doubtful whether we could find our way. By 5 o'clock, however, it lifted. The mornings are so cool that at marching time, 5 o'clock, we have to put on our top-coats. By 9 o'clock the sun waxes so warm that the over-coat is discarded. At 10 o'clock it is warmer; the under-coat is discarded without thanks. At 11 o'clock the vests are divested. At 12 o'clock anything in the shape of clothing is retrained only for the sake of decency. There is usually an alleviating breeze, and as there is no obstruction to the wind, if there is any breeze at all we catch it. The thermometer, too, except occasionally, would not show in the shade any unusual degree of heat. But it must be remembered that we march under a hot, blazing sun on the open plains, with no hill or mountain shade to protect us, and only occasionally along some stream's belt of trees. If we get all the breeze, we get also all the sun. An armament of umbrellas would have been far more useful than our Sharpe or Springfield rifles. The sun has several times endangered the lives of our men, and [the] ambulance has picked up its stricken victims. On the other hand, we have not seen an Indian since we left Fort Rice. If we had, the sight to an Indian of a brigade of hoisted umbrellas would probably have protected us.

MAUVAISES TERRES—THE BAD LANDS OF DAKOTA.

"It is hell with the fires put out," said Gen. Sully, when in 1871,[13] with a wagon train, he tried to go through the "Bad Lands." Metaphorically,

---

[12]Actually Capt. Frederick Benteen (1834–98), who may or may not have deserted Custer on that fateful June 25, 1876, day.
[13]Sully's comment actually was made in 1864, following the Deerkill Mountain battle.

it was a very good description. Factually, it was a great water denudation that left them what they are. As we found, after a more intimate acquaintance, the section of fire is visible here, but it is water mainly that has wrought this devastation and earned for this curious region its bad name. This peculiar character of the country in the North-West is found at intervals in a belt perhaps 600 miles long from north to south, and 200 miles east and west, flanking the valleys of the large rivers and creeks. It is found in the western part of Dakota in extensive tracks running up and down the Little Missouri, the Upper Missouri, the Glendive, and the Yellowstone. It occurs on some of the forks of the Platte. It borders the Black Hills and extends through South-Eastern Montana and Northern Wyoming. These lands are monuments of the wonderful power of water.

Any one who has been to California and seen the effect of hydraulic mining in the hills, washing away, cutting deep gulches through their sides, and pouring the loose earth into the valleys, may form some idea of the Bad Lands on a small scale. In the Eastern States there are no topographical phenomena which can form the basis of a correct comparison. The effect of a heavy flood of rain, which has plowed through a region of yielding sand or clay hills, carving and fluting their sides from brow to base, cutting miniature streams and gulches and transporting hundred[s] of tons of clay and sand into adjacent streams, may suggest something of the physical appearance of an acre of the Bad Lands and the mechanical energy that has wrought it. Multiply this one acre by thousands of acres; instead of one hill have a myriad, stretched over miles of the country [and] . . . looking in the distance like the broken ruins of an aboriginal city.

Imagine this torn, eroded, water-worn, sandy complication of bastions, peaks, hill-cones, 50 to 450 feet high, in the midst of this maze of dry ravine, and you may succeed in forming a notion of the "Bad Lands." It is a weak name, in French and English, and indicates the occasional poverty of each of these tongues. Someday the German, with its unlimited possibility for compounds, may give us a name for these lands which shall do justice to their unmitigated badness. If there are any lands anywhere that can surpass these in barrenness, desolation, and utter unfriendliness, let them be called "worse." If not, this disgrace belongs to Dakota and Montana, and they have a right to the superlative term. The desert wastes of Wyoming and Utah are indeed cheerless and forbidding; but they have the compensation that one can pass over them as fast as he chooses. But once in the bosom of the Bad Lands, you cannot easily escape. The

Atlantic coast has here and there an inhospitable geography. . . . But I cannot recall any place quite as hostile to civilization, so uninviting in all its features as these eroded, water-worn wastes of the North-West. The cañons of the Rocky Mountains and those of the Sierra Nevada are effects of the same hydraulic energy. But there beauty and sublimity have not fled. Man, too, is not excluded. He may dig fortunes from the mountain heart or garner his sheaves on its side. The Bad Lands, on the other hand, are the type of utter unfriendliness. They seem like a remnant of the original chaos that never was finished.

We reached the *Mauvaises Terres* on the 9th of July and camped on a tributary called Davis Creek.[14] On the Plains, wherever the country would permit, our wagon trains usually traveled in three or four columns, with an advance and rear guard of infantry and five companies of cavalry on each flank. This makes a compact column and protection against the Indians is well secured. Coming to the Bad Lands we were obliged to stretch out our train in single file to find our way along its narrow divides and tortuous ravines. If the Indians had any design upon the expedition this was certainly their strong point. Our cavalry, mounted, would have been of little use to us, for it is extremely hard for horses to go up and down these steep hills much faster than a walk. To afford protection against the Indians in this exposed district, Gen. Stanley changed the order of march. The cavalry, instead of marching on the flanks, were placed in front and rear of the train. The infantry were deployed as skirmishers on each flank. Thus every wagon in the train had a guard of three or four men on each side. Reinforcements at any particular point could be centered in a short time.

CHAOTIC MASSES OF ROCKS.

Thus disposed we began our march through the dreaded Bad Lands. Our poor mules suffered not a little in hauling their heavy loads. The pioneers had to be strengthened by heavy details of infantry to reduce the sharpness of the banks, to bridge crossings, and to fill up holes. By any but the scientific party I think the Bad Lands would have been voted out of the route if it had been possible to escape them. They lay at right angles across our path and could not be avoided by a detour. The

---

[14] Named after surveyor Henry C. Davis (1849–1910) by Rosser in 1871. Just south of today's Medora, North Dakota. Today a golf course is located where the creek enters the Little Missouri.

transition from the undulating floor of the plain to the pits and peaks of the Bad Lands was less gradual than I had supposed. From a knoll near our trail a section of this country broke upon us in full view.[15]

It was in the distance as if the cover had been suddenly taken from Herculaneum or Pompeii still covered with volcanic dust and ashes. A little nearer the illusion vanished. The broken walls and minarets turned into clay piles and sand hills, and the streets and roads were the parched alkali beds of the dried-up streams. If the Bad Lands are ever enchanting it is distance which lends that feature to the view. The nearer you approach them the uglier they become. When we thought of passing our wagon train through this labyrinth of gutters and seemingly impassable hills, the remark of Gen. Sully did not seem at all exaggerated. One palliating feature of these almost irredeemable lands is the presence of wood, and of water also, such as it is. A fringe of timber is generally found through the ravines and dotted clumps here and there on the sides of the descent. The bottoms ordinarily yield a growth of good grass. The ravines were nearly all dry. Occasionally they contained a little water, but so strongly impregnated with alkali that it was unfit for use. . . . In some of the large streams where we afterward camped a better quality of water was obtained, but it was all more or less alkaline.

On entering the Bad Lands, one of the first things that attracted our attention was the presence of a loose, cherry-red deposit on the tops of the hills, sometimes extending all the way down to their base. It was also noticeable that the average altitude of the peaks of the hills was much below the level of the plains from which we had descended, as though this singular formation of hills and buttes had been carved out of the plain as an engraving is carved out of a block of wood. Another new feature was the presence of large masses of heavy volcanic slag at various levels on the hillsides and on the bottoms. . . . These masses of slag, looking just like the slag from an iron furnace, were afterwards found capping some of the buttes, though not on the highest ones. We left the column and climbed several of the buttes to examine the red deposit. We found it to be composed of a red clay shale which had seemingly been baked and then broken into small chips. This broken shale resembled brick in color, weight, and texture. The hills presented exactly the appearance that they would if several thousand brick[s] had

---

[15]The location is likely today's rest stop on I-94, immediately north of Sully Springs. See illustration figure 19.

been dumped on their surface and afterward pounded into chips or split into shale. . . . This brick-like shale seemed at first merely a superficial deposit on the crests and sides of the hills. We found none of it in the valleys. If it had been transported it was singular why the crests of the hills should have been favored with a red cap while the intervening lands were almost entirely overlooked. The first thing then to determine was whether this flaming head-dress belonged here or was the gift of some generous glacier or more likely a kindly flood. We laboriously climbed six or seven of these ornamented buttes before we could settle this question. At last a favorable exposure presented itself. On the crown of a hill a great mass of the indurated shale, from 15 to 20 feet thick, lay exposed to such a depth that it was beyond question a part of the formation on which it rested. How deep it extended could not here be determined. . . . On the same butte there were other formations, principally arenaceous, that had evidently been affected by the same heat.

ULTIMATE DESTINY OF THE BAD LANDS.

A variety of this sandstone presented a very peculiar clearage. Instead of splitting off into shale it separated into round, tapering, finger-like lengths, as if a mere bundle of sandstone sticks, or resembling the tapering lengths of chalk prepared for the school-room. Near this ancient geological bakery we also found some sandstone biscuit, more accurately, concretions of sandstone rounded into biscuit shape, and looking as if they might be the dietetic remains of a stone age. In heaviness and indigestibility they would do credit to a New York dining saloon. The exposure referred to above was full of plant remains. The impressions of sedges and plants sometimes determined the point of fracture. . . .

We found here all the required evidences of local volcanic action, not on a grand scale indeed, but similar in character to the action which results in the formation of metallic dykes. Further observation showed that the volcanic level was much lower than that of the highest buttes, which were composed of mingled clay and sand beds, sometimes 40 or 50 feet in thickness, with layers of lignite and traces of iron, but containing not a trace of igneous action.

As we had climbed the hills we had climbed also the steps to a conclusion which seemed logical enough, that the whole level of the Bad Lands showing igneous or volcanic action had at one time been covered by these sandy-clay beds; either that it was covered after igneous action and then denuded, or, what seems more probable, that it was already covered by these beds at the time of the igneous action, and subsequently denuded. The main interest which the Bad Lands had for us was in thus spelling out its history. Water, wind and fire have been the architects of this peculiar devastation. If fire has ceased to be one of the partners, wind, and more especially water, are carrying on the work. The world is not yet finished, or else the streams and flood of the North-West are far too efficiously unmaking it. On every heavy rain that comes, thousands of tons take passage to the Gulf of Mexico and intermediate stations. No soil more readily consents to removal than this. Gen. Rosser informs me that in a single rain-fall he has seen, he estimates, over an inch of the soil taken from these hills. The streams and gullies take their burdens to the rivers, the Yellowstone, the Little Missouri, and the Missouri proper. What is not left by the way is delivered to the Mississippi, to form its bars and delta. Nothing looks so well for the future of Dakota and Montana as the deposition in the Bad Lands to run away.

Published August 5, 1873

Bradley's Diary, July 8–26

JULY 8TH. Marched at 5:30 A.M. Rear guard. Rained from sunrise to 10 o'clock. Crossed Sand Creek at noon and went into camp on Heart river at 3:30 P.M. Afternoon very hot, one man sun struck. Engineers at work ten miles north of our line of march which is their true line. Distance 18½ miles.

JULY 9TH. Marched at 5 A.M. in advance. Bridged Heart River and crossed at 9:30. Heavy fog all the morning. Struck the Bad Lands at 1:30 and camped there miles from point of entrance in a most picturesque spot surrounded by high buttes and ranges affording one of the prettiest views we have had. Somebody ought to give this strip of country a better name than Bad Lands. The soil in the valleys is good, grass and timber abundant, and scenery not surpassed by the [Garden of the Gods] in Colorado. 17 miles.

JULY 10TH. Marched at 5:30 A.M. Battalion deployed on both flanks to cover the train. Passed through magnificent scenery all day. The little valleys between the mountains are pictures of fertility and beauty. Grass very fine and plenty of ash timber. Crossed Davis Creek six times, bridged each crossing. Roads very rough all day. Camped at 6 P.M. on Davis Creek. 10 miles.

JULY 11TH. Marched at 7:30 A.M. on the flanks. Reached the Little Missouri at 8:30. Lay in the hot sun till noon. Making road to the river and treading down quicksand in the stream by crossing Cavalry back and forth. Crossed at 1 P.M. Water thigh deep. Moved out over the cliffs and kept alongside the train which was very much scattered. Afternoon excessively hot and sultry. [*Sentence unclear.*] Marched slow for fear of melting down the men. Trail almost impassable for teams. Camped behind Sentinel Buttes, [or] "the Buttes that look at each other." [They are] Very prominent objects in this broken mountain country. 4½ miles.

JULY 12TH. Marched at 7 A.M. Rear guard. Delayed by waiting for rear of the train which did not get into camp last night. Left the Bad Lands at 10 o'clock and entered a fine rolling country. Cool day and fine roads. Camped at the springs at 4 P.M. 14 miles.

JULY 13TH. Marched at 5 A.M. In advance. Cloudy and cool, fine day for marching. Made three miles 2 hours. Reached Beaver Creek at 11, crossed the Montana line at 10 o'clock.[16] Bridged Beaver [Creek][17] and crossed to the west side and made camp. Rainy afternoon. 15 miles.

JULY 14TH. Marched at 5:30 A.M. Rear guard. Fine cool morning. Passed some rough country, bearing marks of volcanic action. Made camp at 4:30. Dry camp. Crossed Glendive Creek at noon: country rolling and rough. 16 miles.

JULY 15TH. Marched at 5 A.M. In advance. Reached second crossing of Glendive at 10. Very hot day, found poor water in the creek, but was very welcome. Encountered a mountain, we could not turn and had to climb it.[18] Took five hours to get the train up. Fearfully hot all the afternoon and no water. Found a pretty good road on the mountain. Came in sight of the Yellowstone at 5 P.M. and camped on a high bluff overlooking the

---

[16]Actually the wagons had crossed about three miles into Montana the previous day.
[17]Also called Inman Creek for Capt. Henry Inman (1837–99), who later became a well-known western writer.
[18]Instead of following Glendive Creek and making a sharp left on the Yellowstone, the column chose a direct, but far more difficult, march to reach "Stanley's Stockade."

river at 6. The river looked beautiful to us after our long march in the mountains. The stream is full of islands covered with timber and the water very bright and sparkling. 12½ miles.

JULY 16TH. Steamer *Key West* came up today with stores for the Expedition. Lay in camp till the 18th making roads, building stockades and loading trains. Very hot.

JULY 17TH. [No entry]

JULY 18TH. Moved camp about a mile to get on fresh ground.

JULY 19TH. Moved camp down into the valley near the stockade. Steamer bringing up stores from the depot at Glendive Creek.[19] Very hot.

JULY 20TH. The 22nd crossed the Yellowstone to guard the stores being landed on the west side. Baker's train came in from Fort Lincoln today. Had letters from home. Steamer crossing stores and teams to the west side of river. Large details at work on stockade and loading stores. Fearfully hot, 110° in the shade.

JULY 21ST. [No entry.][20]

JULY 22ND. The woods on the [eastern side of the] river got on fire this afternoon, endangering the stores. Turned out 8 companies and put it out after hard work.

JULY 23RD. Very hot and very dry. Dust in clouds; wind and rain at night.

JULY 24TH. Stanley moved his quarters on the river this morning. Moved [my] Battalion down to the river bank and camped in the timber.

JULY 25TH. Broke camp at 5 A.M. and crossed the Yellowstone on the *Key West*. Marched up the valley 5 miles and camped with the balance of the Expedition. Distance marched from the Missouri to the Yellowstone—296 miles.[21]

JULY 26TH. Marched at noon in advance. Engineers on the steamer with Crofton's Battalion, surveying on the eastern side of the river. Camped at 3 P.M. on the bank of the Yellowstone. 5 miles.

---

[19]Initially only one steamer was available.
[20]Does this omission reflect Stanley's drinking, then at its worst?
[21]This figure is far too high. Even leaving from Fort Rice (and not Bismarck), the real number was closer to 230 ± 15.

Bradley to His Wife

No. 16
Camp on the Yellowstone,
July 16th 1873

My darling love:

Here we are at last on the prettiest river I have seen in the west, the picturesque Yellowstone twenty-five days from the Missouri. I tell you, we are glad to get here, for this marks the first and hardest stage of our long campaign. After toiling through the mud of the first two weeks and the rough mountain passes of the Bad Lands, the Yellowstone, as we came in sight of it about 5 [P.M.] of a sweltering hot day, looked with its rapid current, green trees and pretty islands, the most charming thing we had seen. I have written to you about the wet time we had in June and the early part of this month and the difficulty of getting our heavy train over the rivers and sloughs. The Bad lands which we passed through are simply a collection of church steeples and court houses, some of which we had to climb and some we had to go round, by roads of our own making.

Frequently on the march I have had to halt and stack [*unclear*] and set the whole Regiment to building roads for the train to pass. Some of the men grumble a little at this work, others laugh and say they'll take the contract for the Northern Pacific Railroad. We are all well up to this point and we are in extremely good luck so far. I'll tell you why.

We started out with sixty days supplies in our wagons thinking it probable that the Engineers would need that time to survey from the Missouri [to] here, but they have done it in twenty-five days. So that we are that much ahead of the calculations in time and the Engineers promise to complete the survey west of the Yellowstone in thirty or forty days. This will bring us back here by the end of August. . . .

The Steamers have got round safely and the *Key West* is here now, landing supplies and she will ferry us across the river. Another Steamer will be here on our return, to ferry us across again and take back the surplus stores. These Steamers left Fort Lincoln about the time we left Rice. . . .

We struck the Yellowstone between Glendive Creek and Cabin Creek, the country is so rough we cannot get the train down to the mouth of the Glendive where Stanley intended to establish the Depot. So, we shall build the Stockade here, and bring the stores landed at the mouth of the Glendive here by Steamer.

You have no idea of the beauty of some of the scenery we have passed. I have wished a hundred times you were with me to enjoy it. Nothing you have ever seen of mountain views can give you an idea of the Bad Lands. For grandeur and beauty of mountain scenes, they surpass anything I've seen. The Big Horn Canyon[22] doesn't equal this place. . . . There are a great many little valleys in these mountains that are as beautiful as they can be and one of these days, they'll be delightful homes for some good folk who love to live among the mountains. This is a wonderful country in all respects. Our Geologist and Paleontologist attached to the Scientific Corps are enthusiastic about it. The mountains are full of marine and vegetable fossils.[23] I have been out with Townsend and [*unclear*] today and we have got some fine specimens of shells and petrified wood. I shall bring home a box full of these things and I send you in this letter some little clam shells just like the clam we get on the sea coast.

Passing through the Bad Lands, we saw fossilized ferns embedded in the rocks with stems as large as my body and we found the remains of lizards, thirty or forty feet in length. I wish Marsh were with us, for I think he would find a rich field here.[24]

We are very well supplied with stores and the Steamers brought around fresh lots of potatoes and canned vegetables and fruits. Our hunters keep us supplied with game such as Blacktail, Antelope and Mountain Sheep. I haven't tasted beef for a long time and don't know as I'd know the flavor of it. There is no danger of our suffering for food though I do long for some fresh vegetables.

We shall be here for a week or more before starting up the Yellowstone. We have 2,000 mules to shoe before starting and this is no small work, the Engineers are at work at the same time so there is no time lost.[25]

---

[22]Bradley had been stationed at Fort Smith, Montana, on the Bozeman Trail, where it crossed the Big Horn River.

[23]Much of the area Bradley describes is today's Makoshika State Park (Glendive) and is under active paleontological excavation by Montana State University and the Museum of the Rockies.

[24]Othniel C. Marsh (1831–99), a Yale faculty member, was a major nineteenth-century paleontologist. Because of Bradley's New Haven background, it is plausible the two were acquainted.

[25]For those believing Stanley's drinking caused a major delay, this paragraph is particularly informative.

No. 18[26]
Camp on the Yellowstone
August 1st 1873

My dear Wife:

... We left our first camp on the Yellowstone four days ago and marched away from the river to turn [avoid] a canyon through the mountains that we could not pass. After a march of 70 miles through the roughest country I ever saw, we struck the river again tonight. We are in a fine broad valley now and hope to find it good all the way to Pompey's Pillar.

Rosser's Diary, July 8–23

JULY 8TH. One of the teamsters was thrown from his wagon yesterday and the wagon passed over him and killed him. Last night at Sunset we buried him. The ceremony of burial on [the] Prairie is very much like that at sea, but to my mind no[t as] solemn. Very warm day and we marched to Heart River. I effected reconciliation between Custer and Stanley.

JULY 9TH. Left the main column and ran a line across the country and connected on Sully Fork of Heart River and then ran a line up that fork 12 miles. Encamped near the Bad Lands and hope tomorrow to get through to the Little Missouri River. Prospect not promising but will be solved.

JULY 10TH. Very foggy morning and the grass very wet. Ran into the "Mauvaises Terre" and had great difficulty getting down. Finally got on a good stream[27] and got a fair line. Encamped on a very exposed position about a mile from the Little Missouri River.

JULY 11TH. Finished the line on Sully's Creek and made corrections on the survey of 1871. Returned and found the main columns at the crossing of the Little Missouri. Had great difficulty getting up the heights west of the river and the day was very warm.

JULY 12TH. Left camp near the Little Missouri at 6 A.M. after the heavy rain last night. Today it is quite cool, encamped tonight near Sentinel

---

[26]No. 17 is missing.
[27]Railroad Creek, a mostly dry stream, enters the Little Missouri at Medora.

Butte. Hope to reach the Yellowstone in 5 days—then will soon see my darlings.

JULY 13TH. Heavy clouds this A.M. and about Noon began raining and we had a very wet disagreeable P.M. Reached Inman's Fork[28] about 12 Noon and made an indifferent bridge and took most of the day to cross.

JULY 14TH. Crossed over the [Yellowstone–Little Missouri] divide on to the headwaters of Cedar Creek.[29] Road quite good, water scarce and grass indifferent. Encamped on high table but little water. Rained very hard commencing about dark.

JULY 15TH. General Custer with a squadron of cavalry accompanied me over to the Yellowstone to look up the country and to find the boat. Had a very hard ride. Reached the boat 3 P.M. and had it run up the stream about 10 miles. Will accompany the depot to be changed [to] up river.

JULY 16TH. Came up on steamer with General Custer and selected a new depot for supplies and stockade. Had Meigs and Eckelson with me and gave them directions about running the line and will send them tomorrow to resume work on the line from Glendive's Creek up the Yellowstone. Very windy and disagreeable.

JULY 17TH. Sent party down to Glendive's Creek to resume work. They will bivouac tonight and reach camp tomorrow. Gave the party the use of my horse and I am afoot today and have spent the day quietly in camp. Am now getting dreadfully homesick.

JULY 18TH. Camp moved about a half mile. I took a horse from one of the teams and rode down to meet the party. Just as I joined them a large black tail deer ran directly through the lines only a few feet from the group and escaped. Line passed camp 2 miles. Will have to wait now till the boat can be sent up river with escort.

JULY 19TH. Moved camp down under the hill near the river and began making arrangements to ferry [*unclear*] over the river. Our supplies not arrived and the Commanding Officer promises me supplies from his store which will expedite matters very much.

JULY 20TH. Spent the day in fixing up maps. [Wagon] Train got in from

---

[28]Today Wibaux, Montana, on Beaver Creek.
[29]Just south of Glendive Creek.

Fort Lincoln bring[ing] several letters from my darling wife. She does not expect me in August but I will disappoint her and see her sooner than she expects. God bless her and I am *so, so* anxious to see her.

JULY 21ST. Moved my camp from the dust . . . and although I have no guard I feel quite safe. Stanley is very drunk and I fear Custer will arrest him and assume command.[30]

JULY 22ND. Stanley very drunk. . . . [*unclear but not related to Stanley*] The steamer got through landing supplies and I got my camp over the river by hard work, very comfortably and went into camp near the battalion of the 22nd Infantry. Some sign of Buffalo and I hope to see some before I go down river.[31]

JULY 23RD. Had a line run across the river triangulated. Late in P.M. a heavy cloud came up discharging heavy wind and came near blowing over our tents. Covered us with dust. . . .

Phelps's Interview with Custer and Rosser

In the afternoon General Custer came over to our tent, and himself and General Rosser for an hour or more [relived] the incidents of the war, in which they participated on opposite sides. They were fellow cadets at West Point in 1861, Rosser in the first class and Custer in the second, but owing to the necessities of the war, both classes were graduated at the same time. Rosser, however, having been ordered home two or three weeks before the close of the academic year, never received his diploma. Being a Virginian he took service at once in the Confederate army, and before the close of the war he, as well as Custer, rose to the brevet rank of Major General.

Both were cavalry officers and they were pitted against each other on many hard fought battlefields with varying fortunes. They were foeman at Five Forks, the Wilderness, Gettysburg, and nearly all the battlefields of the Old Dominion.[32] Until this occasion they had met but once since the close of the war.

---

[30] And if Custer had arrested Stanley? Any long-term scenario would undoubtedly see Grant making sure Custer was cashiered from the army well before 1876.

[31] During three summers (1871–73) Rosser saw perhaps a dozen buffalo.

[32] Phelps is mistaken. They did not clash at the Wilderness or Gettysburg. However, he omits at least three major battles in the Shenandoah Valley in late 1864.

The interview was exceedingly pleasant and deeply interesting. Their mutual questions and explanations concerning the many combats to which they had been parties, cleared up points which before were mysterious, and afforded an entertainment to their only auditor on this occasion that was well worth the trip to enjoy.

In physique as well as in the general type of their characters, these two noted cavalrymen are as different as could well be conceived. General Custer is of medium stature, with body slightly inclined forward in walking, face spare, nose rather large and pointed, hair light, and hanging in slight curls to the shoulders. On this occasion he wore a black, broad-brimmed, low-crowned hat, a dark blue, short sack-coat, light blue pants, and red shirt with collar turned over the vest. In talking he is intensely earnest and lively, and during this interview he sat leaning forward with his arms crossed and resting on his knees, which were also crossed—not a very soldierly attitude to be sure. His manner is quick and nervous, and somewhat eccentric. As a soldier he is bold, dashing, and efficient. Were he a religionist, you would expect him to be a first class fanatic, for it is impossible for such a man to do anything by halves. He is a good soldier, but has a reputation of being rather reckless in his style of life.

Rosser is a man about six feet two inches in height, broad and erect, with a well knit frame and weighing about 220 pounds. His face is round and full, his hair black, his complexion fair, with a ruddy tinge. He impresses you in his physical presence as a man of great strength and endurance. In conversation his voice is never pitched on a high key, but is deliberate without being slow, energetic without being demonstrative. He is one who is eminently fitted to exert great influence over men and retain a hold on their confidence. He is a growing man. At the close of the war he gracefully accepted the situation; came North and entered upon the work of railway engineering.[33] He is at present the engineer in charge of the Dakota Division of the Northern Pacific, and is personally superintending the operations for the extension of the line to the Yellowstone. He enjoys in a high degree the confidence of the company, and it is safe to affirm that this confidence is in no respect likely to be misplaced.

Source: William Phelps, St. Paul Pioneer, *July 12, 1873.*

---

[33] Not until almost five years after the war was Rosser hired by the Northern Pacific, where his exceptional leadership skills resulted in rapid promotion.

Custer's "Arrest"

THE YELLOWSTONE

Camp Canby, on the Yellowstone, July 23, 1873.

Editor Journal: to say that the subscriber is blue, sick and tired out with this inactivity, is a very mild way of expressing it. The monotony of camp life in this howling wilderness, with never an Indian in sight to scare us a little, is enough to stagnate the energies of a lunatic. I have been here a week, looking for something startling to write you of, but quietude is all I have discovered, with a few exceptions. . . .

General Stanley's command reached the river on the 14th, on a new route, fifteen miles above this point. General Custer, with a battalion of cavalry, who was ordered to hunt us up, arrived here today. He reports everything quiet with the expedition. Not an Indian was seen during the march thus far. The command, however, met with a few minor incidents which served to keep it the least bit excited. It encountered some very terrible storms, some of them accompanied by hail. The stones in one instance were so large that men were pounded black and blue by them. Antelope and deer were killed and mules and cattle stampeded. The hailstones at our camp measured 1⅝ inches in diameter and fell to the depth of a foot or more. They were used by officers in making a decoction [sic] which we termed Glendive cocktails. Two or three men were killed on the trip by falling from their mules and being run over by the wagons.[34]

But the biggest stir created on the trip was the arrest of General Custer by General Stanley. I did not learn the cause of the arrest, but one of Custer's own officers told me that Custer marched[35] in the rear of his command one whole day before he was released. Custer's men denounce his arrest as an act of tyranny, while Stanley's men on the other hand justify their chief in the act. I have no opinion in the matter. It no doubt will be fully explained at the proper time.

Another little breeze, almost equaled to the one just mentioned, was created by an order from General Stanley, directing the commissary of the expedition to destroy all liquors in camp. This order was executed so promptly that there was not a single drink left in camp an hour after it had

---

[34]Actually one, although a second died of disease.

[35]Custer's "arrest" was symbolic, not literal; he rode at the rear of the column but in front of the cattle.

been issued. It was an awful hard blow to the old drinkers, and they were obliged to resort to [another] pain killer, Jamaica ginger, Worcestershire sauce, etc. The sutler to the infantry command lost six barrels, and the cavalry sutler seven. This occurred two days before the command crossed the river, and the day I left there, which was the one following, the topers [being] reduced to vinegar bitters, and they were a forlorn crowd. . . .

Source: Sioux City Daily Journal, *August 9, 1873. The author is unidentified but likely an officer stationed at Fort Buford.*

THE LATE SCANDAL IN THE YELLOWSTONE EXPEDITION

. . . Stanley ordered Custer and his senior subordinate, Captain Hart,[36] to headquarters. It was at guard-mounting about 7 o'clock P.M. I was standing within fifteen feet of Stanley at the time. The two officers rode up to the open space on which the guards were being inspected, dismounted and advanced to Stanley's tent. He was standing alone, outside, in his shirt-sleeves. As the officers approached him, I noticed that no salute passed except between Captain Hart and Stanley. The General then informed Custer that he was in arrest, and directed Captain Hart to take command of the regiment. Salutes were again exchanged between the Captain and Stanley, and the two cavalrymen retired, the whole ceremony not occupying more than half a minute.

The next day Custer marched in the rear of his regiment. The sympathy for him was deep and universal outside of Stanley's headquarters, as far as I had any opportunity to observe. But there was no disguising the fact that Custer had used language disrespectful toward his superior officer in his rejoinder to [a note sent earlier by Stanley]. The provocation, however, was great, as it was felt that Stanley's conduct was unnecessarily irritating in his interference with the domestic concerns of the cavalryman's headquarters.[37]

The kind offices of mutual friends of the two Generals led to something

---

[36]Verling K. Hart (1838–83) transferred to the Fifth Cavalry in 1874 and thus was not at the Little Big Horn.

[37]The issue that brought on the arrest was that Custer's brother-in-law, Lt. James Calhoun, lent a cavalry horse (government property) to his brother Fred, one of the surveyors (a civilian), a technically improper act.

like reconciliation. . . . [However,] it was authoritatively said that the end has not yet been reached, and would not be until after the return of the expedition.

Source: Philadelphia Evening Bulletin, *September 9, 1873. Likely written while the expedition was encamped on the Yellowstone in mid-July. Taken from an undated* Chicago Post *article, author unidentified.*

Konopicky to His Parents

On the Yellowstone River
July 20, 1873

We have reached the river for which our expedition is named—the Yellowstone, but, to be sure, only after long and tiring arches full of the hardships that the desolate land through which we have been wandering can offer. Since I can give you, my Dears, only a small idea of it, I will only say that the country [*unclear*] beyond Fort Rice is nothing more than hilly, partly flat prairie, without trees, water (we had almost nothing else to drink than the rainwater accumulated in deep sand depressions), few living things except antelopes, prairie dogs, toward the end mountain goats (they resemble ibex and are the best-tasting game so far), and a few vultures. By the way, at the Yellowstone we came across traces of buffaloes. Naturally, there is no sign of Indians, that is, wild ones to be seen. We hope to soon make their acquaintance.

Oh, I almost forgot to mention, that there is another kind of animal here—rattlesnakes and in great numbers! Lieutenant [Colonel Frederick] Grant brought me the first of these beasts! I was sitting in the wagon and I took it in my hand, believing it to be dead, when the beast began to rattle. These animals are numerous and grow to a size that I have seldom seen among poisonous snakes. So far nobody has been bitten, except an Indian horse which came up lame.

Before we reached the Yellowstone it took us about two days to cross a sand hill range. I have never seen such things in my life. There were countless earth and sand hills 300 or 400 feet high in picturesque groupings, their domes brick red from previous volcanic fires. Naturally, the passage through this sea of hills was taxing for man and beast. The train of 250 wagons had to negotiate many deep gullies in horrible heat and with almost no water.

Otherwise, I am making the journey as pleasant as I can. In the mornings I usually ride my horse or in the afternoons, when the heat is too much for me, I give my horse to some poor devil of a soldier to ride and I ride in our wagon. We have been on the banks of the Yellowstone for three days and expect to stay here for about another three. It will take that long for the steamboats, which will take the column across the river, to get here from Fort Lincoln.

The Yellowstone is a picturesque, large, fast-flowing stream, that in the area where we are, reminds me of the left bank of the Danube at Nussdorf,[38] with its wooded islands and water filled meadows. . . .

As soon as we get to the left bank of the river, the march will continue. We will follow the Yellowstone to the confluence of the large Rosebud River. From there we will go north through country which has never been explored by white men to the Musselshell River,[39] the final destination of our journey (About 250 English miles from here). We will follow the mentioned river east back to the Yellowstone. If the season is not advanced enough to worry about the rivers freezing up, the Scientific Corps will return home by steamboat via the Yellowstone and the Missouri. If not, we will use the old route to Fort Rice. I am very curious to see [*stops in mid-sentence*].

If the truth were told, I will be happy to see the end of July and August because of the terrible heat. Right now, at two o'clock in the afternoon, it is 108° F. in my tent and I am hardly able to write. It is not possible to exist in the sun.

I have 20 sketches in my portfolio for Washington, which the officers frequently enjoy seeing. Further up the Yellowstone and especially on the Musselshell River, areas that the world will first be able to glimpse through my pencil, I hope I will have more opportunities to sketch than here to fore. General Stanley, the army commander, asked an adjutant to tell me that if I want to stay behind to finish a drawing, I should inform him through Major Lazelle, and he will post soldiers to guard me against the Indians.

Very diligent, or the most diligent of the Scientific Corps are Misters Allen and Bennett, while the geologist [Nettre] and the photographer Pywell have mainly been practicing sleeping. Dr. Nettre is homesick for

---

[38] Located within present-day Vienna, named for its many walnut trees.

[39] At 290 miles long and entirely in Montana, the Musselshell is little more than a stream after late spring thaws.

Munich (he was born in America), where his mother has her permanent home. . . .

I am making daily excursions from the camp . . . of about 5 miles by horse to sketch the landscape and collect. I took one this morning, for example, in the company of Allen, Dr. Nettre, Mr. Barrows, the reporter for the *New York Tribune,* and six of our Indians for protection, all highly armed except me, because my hands were full with sketchbooks, parasol, etc. We took a tiring 2½ hour ride through some high sand hills to some high, very interestingly formed mountains, which I drew. After letting the sun bake me for 1½ hours, I and my companions, who were also suffering from the heat, started back. This time, we could water our overheated horses in the Yellowstone.

Tomorrow we will cross via steamboat to the opposite bank of the Yellowstone and two days later the march resumes again. . . . Please don't worry about not hearing from me for [a long time].

1873 Yellowstone Surveying Expedition
Western Dakota & Montana

Map by Vicki Trego Hill adapted from U.S. War Department maps, 1867 & 1876, courtesy Denver Public Library

Copyright © 2013 by The University of Oklahoma Press.

CHAPTER 6

Late July

The final days at Stanley's Stockade were highlighted by the destruction of the sutlers' whiskey supplies (all sources agree that there were thirteen barrels), Stanley's return to sobriety, and the "disappearance" of Fred Grant. Stanley, barely functional and apparently mortified by his behavior, would have been frantic with worry upon discovering that the son of the president of the United States had gone missing—which was exactly what Grant had intended. Grant, a lieutenant colonel and not averse to drink himself, half-commandeered, half-bluffed his way into procuring a yawl (round-bottomed rowboat) from the steamer *Key West* and, with seven others, merrily floated down some 125 miles of the Yellowstone to Fort Buford. Amazingly, the story was recounted by one of the participants and, while inherently interesting, anticipates Winston Churchill's signature Battle of Britain line by sixty-seven years.

Except for the three companies left to guard Stanley's Stockade, everybody appears pleased to finally depart Stanley's Stockade, the source of few good memories.[1] However, the ramifications of Stanley's drinking can be seen or, more accurately, not seen in Barrows's letters, where his name now appears rarely and the complimentary portrayal of him in earlier letters abruptly terminates.

During this period Rosser returned east. While he was no longer on the survey, by mid-August he was back in New York where his enthusiasm "sold" the Northern Pacific's Board of Directors on building the railroad from Bismarck to the Yellowstone at Glendive Creek. In another

---

[1] The stockade was left under the command of the Seventh Cavalry's Capt. Frederick Benteen, a studied insult as it gave Benteen no chance to participate in the expected Indian fighting.

"what if," had railroad construction begun in 1874 and with fighting highly likely, Sheridan undoubtedly would have used Custer to guard the builders. Thus the 1874 Black Hills Expedition and the battle at the Little Big Horn might never have taken place.

Chapter 6 combines two of Barrows's shorter letters to the *Tribune*, written July 24 and 27, Grant's ride down the Yellowstone, and the diaries and letters of Bradley, Meigs, and Rosser.

That same July, as the survey moved forward, the popular writer Charles L. Brace (1826–90) visited Bismarck, possibly in preparation for another of his best-selling travel books. Brace represented Anglo-American aristocracy; his family (De Bracy) had Norman roots, a great-grandfather signed the Declaration of Independence, and he was a product of Yale and its divinity school. Brace's published books included his adventures (more than he'd wished at times) in Great Britain, Germany, Hungary, Norway, Sweden, and California. However, he also had time to create the Children's Aid Society, take an active role in penal reform, and write about urban poverty in Manhattan's dangerous Five Corners neighborhood, where he chose to make his home. One of his letters concerning the "New North-West" describes wide-open Bismarck and is included here.

Barrows to the Tribune

YELLOWSTONE EXPEDITION. PART I

AT THE BANKS OF THE YELLOWSTONE.

THE BUFFALO IS A GUIDE—FIRST GLIMPSE OF THE YELLOWSTONE—SUCCESSFUL VOYAGE OF THE KEY WEST—IN CAMP ON THE WEST SHORE OF THE RIVER.

YELLOWSTONE RIVER, MONTANA, JULY 24.—The buffalo is a great engineer. For the Western country he is the best guide that one can have. It was a buffalo that led us the last 80 miles of our journey to the Yellowstone. I do not wish to detract from old Clemmo, whose knowledge and experience as a guide are invaluable. It is another proof of his fitness that he had wit enough to follow the path which the buffalo had made. After getting out of the Bad Lands of the Little Missouri, we soon struck the Yellowstone slope. Along the valley of the Yellowstone there is another break of Bad Lands almost if not quite as difficult of passage as those flanking the Little Missouri. Clemmo and a company of cavalry were sent to make a reconnaissance and on their return reported that the roads were very bad. Previous to this we had followed mainly the trail of Gen. Whistler, who commanded in the survey of 1871. Gen. Stanley determined now to find a new road. His reason for this was that Gen. Whistler's trail left the divide and broke off into the Bad Lands, through which, on one day's march, he was obliged to make 22 crossings. The only way to travel easily through the country is to keep on the top of a divide or ridge, and never leave it unless necessity compels. It is shorter and easier in the long run to keep on the divide, even if it takes you considerably away from a direct line. It is this method the buffalo follows. A certain topographical instinct keeps him on the lowest divides; somehow he succeeds in finding them easier than anybody else. By following the divide between Glendive and Cedar Creek we were enabled to make 17 miles. We found the road much better than we had expected. In the course of our march we came upon two villages of prairie dogs. The whole surface of the plain was covered with their little mounds . . . and the whole community kept up a most courteous barking.

WITHIN ONE DAY'S MARCH OF THE RIVER.

The only thing of geological interest we met during the day was a small sand hill not over 20 feet in height, standing isolated on the plain. . . . We camped that night only one day's march from the Yellowstone. We set out the next day, July 15, computing the distance to the river at 12 miles. The great point of anxiety was whether we should find a boat on the Yellowstone at the mouth of the Glendive. Three steamers had left Fort Rice before our departure, bound for the Yellowstone at this point, with provisions for the command. One of these steamers, the *Key West*, we hoped to find to ferry us over. It was a question whether with her heavy load she would be able to proceed so far up the river. Gen. Forsyth's success[2] [in May 1873] going up to the mouth of Powder River in this same vessel was our ground for trust. If the boat had not arrived, we should be obliged to send to Fort Buford, 100 miles, for supplies. To determine the matter as soon as possible, the cavalry was ordered to move on in advance at five o'clock in the morning and search for the boat. The cavalry started across the country. The train, under guard of the infantry, was obliged to be more particular about its road.

CAUSES OF THE DISAPPEARANCE OF THE BUFFALO.

At the start a view of the section of country over which we were to pass was rather discouraging. The Bad Lands seemed to lie before us in every direction. But we still had confidence in Clemmo, and Clemmo, though he said nothing about it, evidently confided much in the buffalo guide that had passed over this country two or three years ago. We noticed that he kept close to the trail and that the trail led due west. Though from Fort Rice to the Yellowstone we have not seen a single buffalo, the marks of his range are everywhere visible. The hills are covered with his wallows and trails. Occasionally buffalo chips two or three years old are found along them. What has so completely driven the buffaloes from a country where they once ranged in such vast numbers? A few years ago, along the Missouri clear up to Fort Buford and beyond, the plains were frequently black with them. . . . I have asked the opinion of many experienced hunters and army officers on the point, but do not find any

---

[2] Maj. George A. "Sandy" Forsyth (1837–1915) was the hero of the Beecher's Island Indian fight a few years earlier. A member of Sheridan's staff, he is sometimes confused with but not related to Lt. Col. James W. Forsyth (1837–1906), also on Sheridan's staff.

general agreement as to the cause.[3] Some refer it to the Sioux. But the Sioux have roamed and hunted this country for many years, and still the buffalo was seen in abundance. Others more reasonably refer it to the vast fires which have ravaged the country, burning off the grass and driving the buffalo further west for good pasturage. Indirectly the Sioux may be responsible for this prairie arson. Excessive drouth, which, according to the Indians, has been more severe in the last ten years than at any other time, is another cause which, combined with the previous one, would cause the buffalo to seek a more congenial range. The buffalo must have water. When the Indian and trapper, far from their accustomed haunts, long for water they invariable follow a buffalo trail, assured that if water is to be had they will find it. Perhaps, once again, should a few successive Summers be as moist as this, Dakota may once more become the buffalo's home. Now it is only a vast buffalo cemetery. His whitened skulls and bones, scattered far and near across the plain, find no grave, but in a few years from now will not lack an epitaph. The buffalo is dying out.

FIRST VIEW OF THE YELLOWSTONE.

Our march through the morning was made without difficulty until we came to a small stream which gave us a little trouble. Beyond this a short distance we reached the foot of a very high and steep ridge. Old Clemmo insisted that if we did not take it we should not reach the Yellowstone that night; but the valley to the right looked more inviting. An inexperienced man would have refused the hill and accepted the valley. Gen. Stanley knew better. He sided with Clemmo; it was better to spend four hours in getting the train up to the top than to trust to the spacious valley, which would soon inextricably involve us in the Bad Lands. The result proved this view correct. We had to double our pioneers and double our mules on the wagons. But once on the top of this divide, there was no more trouble. After another hour's march we struck a beautiful broad plateau, as level as a floor, two or three miles in length, as many broad, and stretching nearly to the river. As we entered on this broad plateau, our eyes were gladdened by a beautiful view of the Yellowstone, glistening in the sunshine, with its green relief of cottonwood thickly shading its numerous isles. It was the most gladdening

---

[3]There is still no uniform agreement. One likely cause was disease spread by escaped hogs or cattle. Another was the killing of three-year-old females, whose hides were the ideal. As the Indian's desire for arms (and liquor) increased, so did the rate of his killing.

sight since we left the Missouri. We had marched all day in a blistering heat, so extreme that several men had given out. The sum of all human wants seemed comprised in water and shade. About a mile from the river, the plateau shelved off into Bad Land, and finding no way down to the river, we camped on this table, and sent our animals to water.

ARRIVAL OF THE KEY WEST.

The cavalry meanwhile had had a very hard march. Striking across country they became involved in the *mauvaises terres* which in many places are almost impassable for horse, compelling severe climbing and frequent detours. After marching until late in the afternoon without any signs of the boat, they were seriously considering [*line missing*] . . . when the mark of a horse shoe was discovered in the green. It was clearly not the track of an Indian for his pony is never shod. A close examination showed that it was the print of a United States shoe. United States troops were somewhere in the vicinity. Gen. Custer concluded that it was the mark of some of the troops brought up by the boat who had probably camped near by. The march was pursued and in a short time they had the great satisfaction of finding the steamer *Key West*, Capt. Grant Marsh, which reached Glendive Creek on the 1st of July and made a return trip to Fort Buford for additional stores. Seeing that it was impossible for the wagon train to make its way [to] Glendive [Creek], Gen. Custer wisely determined to have the boat go up the river. He embarked his men and horses and proceeded ten miles up the river, where he disembarked, picketed his horses and bivouacked his men. . . . On the following day the news of the arrival was conveyed to Gen. Stanley.

The *Key West* had brought up a large quantity of stores for the expedition, in charge of Lieut. R. M. Thorne, Regimental Quartermaster of the 22d infantry. The steamers *Peninah*[4] and *Far West* had also brought a cargo, and the whole had been left at Glendive Creek, under guard of three companies of the 6th infantry, from Fort Bedford, commanded by Capt. Hawkins. As it was impossible for the expedition to [reach] Glendive [Creek], it was necessary to reship the stores up the river. This took four or five days. Meanwhile Gen. Stanley had succeeded in making a road down to the river, and camped on the flat.

---

[4] The *Peninah* was a 287-ton sternwheeler, wrecked in 1875.

IN CAMP ON THE YELLOWSTONE.

On the 20th inst., Capt. Baker, Chief Quartermaster, who, on the 3d of July, had been sent to Fort Lincoln with the train of 50 wagons, reached our camp on the Yellowstone. He had come through in 12 days. He had found two of the notes which we had placed on sticks and left on our trail, informing him of our rapid advance and of the men left behind as additional guard. At the proper time he fell in with Capt. Pearson and his men, and the whole party advanced rapidly to the Yellowstone. On their way out the party had seen 11 Indians, who got away as quickly as possible when they discovered how well the train was supported. An amusing incident was connected with their approach to Fort Lincoln. The commander was not expecting them, and on sighting the cavalry in the distance, supposed they were so many Indians, who had captured the train and were advancing to attack the fort. The long-roll was sounded and the guard brought out. The soldiers were all primed for a fight. The mistake, however, was soon discovered, and the would-not-be Indians were accorded a warm welcome.

As soon as the steamer for *Key West* had completed the shipment of stores to our camp, she was employed to ferry the command over the river with sufficient forage and commissary stores for a trip of 50 days. The remainder of the stores greatly in excess of our needs were left on the east side of the Yellowstone. A suitable stockade was erected. Captain [Benteen] . . . has been placed in command of the depot. His troops consist of 40 infantry and 119 cavalry—in all eight officers and 159 men. Capt. [Benteen] will remain here till the return of the expedition from its explorations in Montana, which it is expected will take 50 days.

We are now in camp on the west side of the Yellowstone, expecting to start in a day or two. Meanwhile the boat will proceed up the river 25 miles with engineers, and an escort under command of Major Crofton of the 17th Infantry.[5] This will take about two days. The boat will then return to Glendive Creek, take on the 6th Infantry, and go down the river.

Published August 19, 1873

---

[5]Maj. Robert E. A. Crofton (1834–98) was on Stanley's 1872 survey. He married into the du Pont family and was a Civil War veteran held in high regard by Sheridan. He was twice tried in civil courts for murder, in at least one case for killing a man he had cuckolded. In the 1890s Crofton was twice shot by soldiers (one an officer) for making (successful?) advances to their wives. Students of American labor history know him from his defeat of Pullman strikers in Chicago.

THE YELLOWSTONE EXPEDITION. PART 2

A STEAMBOAT TRIP ON THE YELLOWSTONE— THE GOVERNMENT PAYS FIFTEEN THOUSAND DOLLARS A MONTH FOR A STEAMBOAT—SCENES ALONG THE RIVER—UNCEREMONIOUS DEPARTURE OF FRED GRANT.

ON THE YELLOWSTONE, JULY 27.—It seems not a little singular and perhaps reflects somewhat on American enterprise that one of our largest and most beautiful rivers, nearly 300 miles long[6] and navigable for 200 miles from its mouth, should remain entirely unexplored by large steamers until the year 1873. It is 70 years since Lewis and Clark explored the Missouri to the Rocky Mountains and returned by way of the Yellowstone, sailing in small boats from its head waters to its mouth at the Missouri. Since that time hundreds of steamers in going up the Missouri have passed the mouth of the Yellowstone, but none of them have ventured to turn their prows up the river until this Spring. The fact that there are no settlements on its banks except the camps of hostile Indians, and its supposed unfitness for navigation, which had come to be accepted as a fact, no doubt deterred Missouri captains from risking their boats in its waters. It is strange, however, that the Government itself did not undertake to settle the fact of its navigability until this Spring, when Gen. Forsyth, under Gen. Sheridan's direction, made a successful attempt to ascend.

FIFTEEN THOUSAND DOLLARS A MONTH FOR A STEAMBOAT.

Sailing up the river, as I now am, on the very boat that first made the trip, to all appearances we are navigating a river as large as the Ohio, and greatly resembling that stream in its flow and aspect between Pittsburgh and Wheeling. It is difficult to realize that this is only the second trip of a steamboat on this river. Finding that the engineers had two days' work to do on the river above our camp, in which it would be necessary

---

[6]*Merriam Webster's Geographical Dictionary* (2001) gives the Yellowstone's length as 671 miles. At high water, it was navigable by smaller steamers to present-day Billings, some 350 miles.

to take the *Key West*, I gladly availed myself of the opportunity to see 25 miles of the river above a point 180 miles from its mouth. The *Key West*, which has made this river trip so successfully, is one of the largest boats of the Missouri, measuring 200 feet in length and 33 feet in breadth. Like all Western river steamers, she is of light draft, drawing only two feet, and has a sternwheel. Capt. Marsh, her commander, is an experienced navigator, and has turned this new venture into considerable profit. Five hundred dollars a day in Government employ for every day he remains in the Yellowstone, whether active or idle, is no mean compensation. Thirty days on this river in the service of this expedition makes a total of $15,000—a profit which, considering the risk, is rather large, nearly enough to pay for his boat. . . .

SCENES ALONG THE RIVER.

From measurements made by Gen. Rosser, it is found that the average width of the Yellowstone is 1,200 feet—its rate about six miles an hour. It is 700 feet above the Missouri at Bismarck. It has a fall of about two feet to the mile which gives it its rapid current. Imagine a steamboat sailing on a river nine times as high as [New York's] Trinity steeple, and you have our height above New-York bay—about 2,300 feet. The average depth at this point is about three and a half feet.

Our boat proceeded up the river without much difficulty, though it is a hard tug against the current. Four miles an hour against the stream is the maximum speed. With the current we can make 16 to 18 miles.

When ascending the river with a heavy cargo, the current is so strong that in some places it would be necessary to "warp," as is frequently done on the Upper Missouri. "Warping" consists in taking a line from a pair of steam windlasses in the bow of the boat, and making it fast on shore. The windlasses are then wound up, and the power acquired in this way, in addition to full steam on the paddles, usually carries the boat up against an obstinate set in the current. With our light load, we met no place that could not be stemmed with the ordinary power of our engines, though in rounding the numerous islands a little backing and veering was sometimes necessary.

The scenery on the Yellowstone, or at least on the 25 miles comprised in our trip, does not differ materially from that on the Upper Missouri or along the Little Missouri. From the dull yellow bluffs which line its banks, it gets its name, and these are often enough repeated to justify the appellation. The average height of the river bank ranges from six to twelve feet. The Yellow Bluffs never come boldly and squarely to the front like the Hudson Palisades, but being composed of soft clays and sands, have been seamed and sloped to the water, preventing the usual phenomena of the Mauvaises Terres. The monotony which would be occasioned by a succession of these soft, yielding, treacherous clays . . . is varied by frequent low green shelves and table-lands, the bluffs kindly receding a mile or more from the river, or entirely disappearing for the time. The change to a green, grassy flat after a succession of these barren, treeless bluffs is very pleasing. The stratification of these bluffs is horizontal and very regular. Dark layers of lignite are sometimes interposed. For two miles of our trip there was a fine exposure of coal about three feet in thickness. The timber is confined mainly to the islands. On these there is usually a fine growth of cottonwood. Large numbers of wild geese frequent the river, and on a requisition from two or three of our best shots several were transferred to the steamboat table, which, considering our distance from civilization, preserves a remarkable concord with our appetites.

DEPARTURE OF LIEUT. [COL.] FRED GRANT.

The trip up the river and back was made without molestation. The Indian, like the buffalo, had left us his only grave. The forests showed, however, that "Mr. Lo," as the soldiers familiarly call him, had been here very recently. The *Key West* returned to Gen. Stanley's camp this evening. She goes down the river to-morrow. Gen. Rosser and several others return by her. Fred Grant took rather unceremonious leave a few days since on a yawl which was sent down to Fort Buford with orders[7] for the steamboat *Josephine*. It was a rash undertaking, and was done without the knowledge of Gen. Stanley, who cannot be held responsible for impeding a valuable life.

Lieut. Thorn, who had charge of the government of the *Key West*, was accompanied by his wife and three little children. Mrs. Thorn has the honor of being the first white woman who ever sailed up the Yellowstone. Many an officer in camp who had vowed not to shave before he got back suddenly reconsidered his determination before going to the boat, and the vanity of boot-blacking was once more introduced.

Published August 23, 1873

---

[7]The "orders" were a polite, face-saving fiction.

"Grant's Perilous Ride"

DOWN THE YELLOWSTONE IN OPEN YAWL
WITH THE PRESIDENT'S SON—A FEARFUL STORM AND
A COUPLE OF SHIPWRECKS—WHAT SORT OF
FELLOW FRED GRANT IS.

Among the many startling incidents and hair-breadth escapes born of the military expedition now wending its slow and tedious course through the Bad Lands of Montana, there is one which I think will prove of interest to the general reader. At least the participants in the adventure will always think of it with feelings of awe, and now that I am safely in the offices of the steamer *Key West*, homeward bound, I am surprised to find the hair still vegetating on the top of my head.

The adventure of which I write consisted of a trip down the Yellowstone in an open yawl, from Gen. Stanley's camp to Fort Buford, a distance of 150 miles, the great portion of which was through the heart of Sitting Bull's hunting grounds.

We left Stanley's camp on the afternoon of the 23rd of July. Our party was made up of [Lt.] Col. Fred Grant,[8] Capt. W. H. H. Crowell,[9] "Harve"—which is his front name, his surname I never knew—the subscriber, and six soldiers who were detailed by Major Baker, Quartermaster of the Expedition. The trip was a very dangerous one—a fact which we all fully realized; but we desired a change; anything for a change. The death of one or two of the party was preferable to the dull, stagnating monotony of our camp life.

We swung out into the rapid current just after breakfast and ran down to Camp Canby, fifteen miles below, where we took a lunch and remained a couple of hours. Capt. Crowell joined us here. We pushed away from the shore at Canby between 4 and 5 o'clock, and ran about 50 miles before trouble commenced.

The day was calm and warm; not a breath of wind was astir to ripple the placid bosom of the river, and the only thing to excite our attention was the occasional appearance of elk, deer, and smaller game. One very large grizzly bear swaggered out from the jungle within easy rifle range, but we restrained from firing for fear of attracting the attention of some lurking

---

[8] Only Custer was "in" on Grant's "practical joke."
[9] Sixth Infantry Capt. William H. H. Crowell (ca. 1840–19?), was a Civil War enlistee who retired with the rank of major in 1900.

band of Indians which we felt sure must be roaming over that country; and just before darkness spread its impenetrable gloom over the beautiful prospects of that wild river, a superannuated government mule, which had been left by the expedition, came down to the water's edge and drank.[10] Darkness was now fast obliterating all objects on either shore, and our danger of being gobbled by Indians was accordingly growing less. The atmosphere still remained motionless and little did we think that from "such calm eve such awful morn could rise." We floated with nothing to disturb the quietude, save the dip of our oars into the tranquil waters.

This state of things was continued until about 10 o'clock P.M. when a sudden change came over the spirit of our dreams. Thick, lowering clouds rushed suddenly up the western horizon and an impenetrable blackness gathered round about us with an astounding rapidity.[11] An awe inspiring roar came over the prairies from out those clouds, and before we could get our boat to the shore, it was caught up by a tornado and tossed about like a thistle down on a summer breeze.

It was a terrific storm. Great trees upon the shores bent over and lashed the turbulent waters with their branches, and the rain poured down in torrents. The darkness was simply thick and the lightning terrifying. The pilot ordered the oarsmen to pull with their utmost strength, but before they recovered their presence of mind our boat was dodging about and pounding against an acre of large snags. At this instant two of the men sprang overboard, one of them with the boat's "painter," who succeeded in snubbing her to a snag. The first strong breath of the gale was now exhausted, and we all felt we had received the benefit of the "big scratch," that we had not been capsized and drowned.

Still there was plenty of confusion left among the members of the "outfit." The subscriber got somehow mixed up with the oarsmen, lost his hat overboard, and finally got down in the bottom of the boat with the whole frantic rabble tramping upon him. *So much confusion was never before gotten up by so small a crowd in such a brief lapse of time* [emphasis added]. Everybody was captain just before the storm struck us, but now there was not a captain in the crowd. The captains were all looking out for No. 1.

As soon as the boat was made secure we settled ourselves down in the bow to await the dawn of day. Sleep was out of the question. Drenched

---

[10]The mule wasn't the only one drinking.

[11]An 1873 almanac indicates that by this time the moon had set.

with rain, the danger of our situation and the excitement of our "shipwreck" kept us wide awake.

In our anxiety to get out of Indian country, we let go the "painter" before daylight, and pulled down into a rapid chute, and ran over a rocky [rapids] and were the second time "shipwrecked." After tugging away for some hours we again got afloat, and in due time reached Fort Buford much exhausted, but mighty thankful for our safe deliverance from hostile Indians, dangerous rapids, and tornadoes. . . .

I desire to answer the question, repeatedly asked to me, "What kind of fellow is Fred Grant?" To be brief, he is social, well bred, well educated, and perfectly gentlemanly young man. He is lively, yet modest; he puts on no "airs," and does not presume upon his father's exalted position to bolster himself up. He readily adapts himself to the company he is in, and does it with heartiness which invites confidence and respect. The anti-administration may call him stupid, boorish, dissipated, etc., but when they make these statements I for one will fully realize that they either know nothing about what they are saying or else are stooping to malicious slander.

Source: Sioux City Daily Journal, *August 17, 1873.*

Meigs to His Parents

Camp near Powder River
July 25, 1873

My dear Father and Mother:

I fear I shall be unable to chronicle the events that have occurred since I last wrote with anything like minuteness. Suffice to say we have arrived here after a long march and are now only waiting for the crossing of our clumsy train to wind our way rejoicingly southward towards our goal "Pompey's Pillar."

We have accomplished very substantial results as far as our R.R. is concerned by the discovery of a route which will shorten the distance between the Missouri and the Yellowstone some twenty-two miles at least, which at the rate of $50,000 per mile makes a saving to the company of $1,100,000, no mean result of a three week survey.[12] Our train

---

[12] The estimated $50,000 per mile figure is quite high, although the number of miles to be eliminated was accurate.

delays us very much it is so large that every small stream crossing or mud hole causes a stoppage of some two or three hours for we have to march pretty well closed up and the advance of the train has perpetually to wait for the rear whenever an obstruction occurs on the road.

Mr. Lo[13] has been kind enough to keep himself out of sight almost entirely since our scrimmage at the Missouri River which was with some Grand River Indians who are ostensibly at peace with the white man. We have scoured the country far and near and found no sign of Indians. . . . [*Remainder unclear*].

This country is so dry and the grass so poor that we find but little game of any kind. Buffalo there are none nor have been for some time, though we see an occasional black tailed deer or Rocky Mountain sheep. Gen. Rosser has secured the stuffed head of one. . . .

I enclose an unfortunate sketch which I attempted of some bad land bluffs below our camp which will however assist you to some conception of the scenery. We don't find the Yellowstone the [placid] stream of 1871. It is up and riled this year and its turbid waters do like the Tiber shakes its tawny mane and boil around the snags at a speed of six or ten miles an hour. The woods on the low points are lovely. The timber is cotton-wood of no great size but seems to have been planted by some landscape artist of taste and offer, destitute as they are of underbrush, the most charming vistas of long perspective, only needing the deer peacefully browsing to make one think himself again on the Nieces [*sic*] or Llano of Texas.

The soil of the valley is here very poor. The right bank of the river is bounded by high bluffs of Mauvaises Terres which often overhang and encroach upon the stream and rise to a height of some two hundred to three hundred feet. These Mauvaises Terres are most indescribably rough and offer a horseman about such a maze of fissures and impassable ravines and ridges as the Mur de Slace[14] at the Montauverti. I rode ahead of the train with Gen. Rosser and a squadron of cavalry some thirty miles and the climbing [by] these horses would astonish the natives. We climbed over ridges or hog backs which it would be supposed a mountain goat could scarcely scale and indeed it was a risky thing for the horses as a single mishap would have been fatal.

---

[13]"Mr. Lo" was slang for an Indian. If you were "cool," this was the expression you used. Nobody really knew where it came from, although one explanation was that it came from "Lo [as in 'see'], the poor Indian."

[14]Did Meigs mean *Glace*, and was he referring to an area in the Pyrenees?

We rode on down to the mouth of Glendive's Creek and found the steamer waiting for us with provisions and forage in abundance. That was some twelve days ago. As the country between was impassable however we had to have everything reshipped and brought twelve miles up to this point where a stockade has been built and they are going to leave two companies to guard the extra supplies for our return. The ferrying and the moving of supplies has consumed much time and we are anxious to get away though our camp here is most pleasant. Every evening we step down to the river about a hundred feet away and bathe. The weather has been intensely warm and three times a day has seemed none too often for us to enjoy the delights of swimming and cooling ourselves in the delightfully tempered water. What do you think of having the thermometer at 103° to 109° for four or five consecutive days? Is it not fair to say though that the heat was comparable to what it would have been in Washington as the dry air carried off the perspiration rapidly and kept one completely cool.

We have a pleasant party. Eckelson you have heard me speak of before. He is a noble fellow and a most general favorite, always full of fun and high spirits, energetic and bold, the very man for such work as ours. He will command our party when Gen. Rosser leaves which the Gen. expects to do as soon as they get through with the steamer now employed ferrying.

I think we may move from here tomorrow but cannot say yet. We have about five hundred miles of line to run yet and think we can average ten miles per day. If so we will make the round trip in about fifty days. We go up the Yellowstone to Pompey's Pillar, then north to the Musselshell Valley and part way by it and part across the divide back to Glendive's Creek.

We are feeding seven pounds of oats to the horses [daily] but they do well on the grass we have had so far. . . . We have a hunter with the outfit who kills a great deal of game. Out of thirteen sheep he has seen, he has killed six since we struck the Yellowstone Valley, and antelope five or six per day often. There is comparatively little game here though. It is curious the total change in the country from the east to the west side of the valley. East the bluffs crowd the river and are steep, precipitous badlands. West the valley slopes gently upward and is intersected by small valleys abounding in springs and fringed with nice ash and cottonwood trees. There is much more gravel in this soil which allows of the water soaking in and makes the country much less dry. In spite of it all however without manure and irrigation I doubt if any of the valley here is cultivable. We hope to strike better country somewhere above here.

We are now having delightful weather and looking forward to our trip up the river. The steamer will go with us as far as Powder River and we will live on board perhaps until we make our connection with last year's line. But I must close in haste. . . .

Bradley's Diary, July 27–August 3

JULY 27TH. Lay in camp all day, opposite the mouth of Cabin Creek, waiting for the Engineers. Very windy and dusty.

JULY 28TH. Marched at 6 A.M. Rear guard. Left the river and marched northwest to turn a line of bluffs butting on the river which we could not pass. Country rolling and rather poor. Camped at 3 P.M. on unknown creek. Distance 15 miles.

JULY 29TH. Marched at 5 A.M. In advance. Country very rough and poor, fully as bad as the Bitter Creek country in Wyoming.[15] Day cool and showery. Camped at 3 P.M. on Bad Route Creek.[16] Distance 15 miles. No game in the country, have not seen an antelope since we left the Yellowstone. No water on the route and grass very poor.

JULY 30TH. Marched at 5:45 A.M. Rear guard. Rained in the night and the morning fine and cool. Country rolling and rough, very rocky in places, but good trail for the train. Custer ahead with a squadron of Cavalry trying to find road to the Yellowstone. In the afternoon struck a piece of Bad Lands we could not get through. Backed out and went round them and reached a camping place in among the hills at 7 P.M. 13 hours on the road and very tired. Very little wood, water or grass. Distance 18 miles.

JULY 31ST. Marched at 5 A.M. In advance. Had to halt and build road the first mile. Country very rough and almost impassable. Looks as though it had been burned out at some remote period; rocks and hills bear the marks of fire. Strong working parties out all day building road all the forenoon. Struck Custer's Creek[17] at noon and followed it. Crossed it 16 times in afternoon, found large veins of fine coal along the Creek.

---

[15] East of Rock Springs, Wyo.
[16] A half dozen miles north of Fallon, Mont.
[17] Custer Creek enters the Yellowstone from the north some four miles above the Powder River.

Reached the Yellowstone at 5 p.m. Found a fine valley stretching out to the west and looking very pleasant. Found the steamer *Josephine* here waiting for us, having run Wolf Rapids. She brought us letters from home. They were very timely for we cannot get another mail for 50 days or until we return to this point. . . . Distance 14 miles.

AUGUST 1ST. Lay in camp all day, fitting up for the long march up the Yellowstone and return. Sent letter home by the *Josephine* which leaves for Fort Lincoln in the morning. Gen. Rosser of the Engineers . . . return[s] on the steamer.

AUGUST 2ND. Marched at 5:45 A.M. Rear guard. Fine morning. Left the river to avoid bluffs that we could not pass and returned to the river again in about 4 miles march. After 4 or 5 miles march in the valley with numerous dry runs and arroyos to cross, men obliged to leave it again and take to the hills. Found country very rough and after a hard day's work, reached the river again and camped at 7:30 P.M. Distance 12 miles.

[SUNDAY] AUGUST 3RD. Marched at 5 A.M. in advance. Kept up the valley till 10 then made a detour to the north. Returning to the river and camping near it on Sunday Creek at 11:30 A.M. Country and grass very good. Distance 11 miles. Very warm weather.

Rosser's Diary, July 24–August 19

JULY 24TH. Took a squadron of cavalry and rode out [toward] the [*illegible*] River about 20 miles to examine the country. Had a very hard and disagreeable ride and came back tired and quite sick. Can't say when the steamer will leave and I am getting out of patience.

JULY 25TH. All work that we can do in this vicinity is done and we have been idle today. Took a long swim of about 3 miles down the Yellowstone. Had my clothes sent down by wagon and teamster lost my hat.

JULY 26TH. Shuck camp 6 A.M. and took the party to the steamboat landing. Ready to go aboard as soon as the horses were ferried across the river. Went aboard 10 A.M. Ran up the river to the line [*unclear*] and escort off. Then ran 5 miles west of Cabin Creek. Party ran 8 miles and walked 1½ to the boat. I am quite sick today. Twice I had a chill. Too much bathing in the Yellowstone.

JULY 27TH. Ran the line up the right bank of the Yellowstone and connected at 4:30 P.M. with the survey of '72 at O'Fallon's Creek. I robbed an Indian grave at this point of a few trinkets. Steamer returned to Stanley's camp, where I took leave of the party. All were well.

JULY 28TH. This A.M. at 5 o'clock we leave expedition and start for home. I must confess that I leave the party with some regrets, but when I realize that I am going home to meet my dear wife and children I cast them aside cheerfully.

JULY 29TH. Stopped at the stockade to take on a few sick men, then ran to Glendive's Creek to take on the 3 companies of the 6th U.S. Infantry. Found several out hunting in A.M. Met the *Josephine* at sunset [*unclear*].

JULY 30TH. Ran to Buford where we part [*sic*] of the detachment of the 6th Infantry. Called on General and Mrs. Hazen.[18] Very pleasurable visit. [*unclear*] and in P.M. we started down river. Mosquitoes worse than I ever saw them.

JULY 31ST. Lay last night at [Fort] Berthold and went up on the hill and examined the Indian village. Very curious and interesting indeed. Ran down the river and reached Bismarck 6 P.M. Telegraphed Mr. Roberts, Mr. Mead and my precious wife.

AUGUST 1ST. Visit to Fort Lincoln with Commissioner Smith and party and witnessed the friendly Indians. Left Bismarck in the P.M. and came through to Fargo. Had a pleasant time as we had a large party [going] east and two Palace cars.

AUGUST 2ND. Ran through on the regular train to the junction where we lay over to await the down train from Duluth which we took and ran the party to St. Paul and went to my home in Minneapolis.

AUGUST 3RD. Spent the day quietly at home with very dear wife and children. Went to church in the evening.

AUGUST 4TH AND 5TH. At home at work on my report of Yellowstone survey.

---

[18] Col. William B. Hazen (1830–87), West Point '55, commander of Fort Buford. He and Sheridan quarreled over who reached Missionary Ridge's crest first at 1863's battle of Chattanooga. Hazen later feuded with Custer and Stanley. Hazen and Rosser's mutual antipathy for Sheridan would have made the evening enjoyable for both.

AUGUST 6TH. Went to St. Paul and saw General Terry. Mr. Mead had gone up on the road and I did not see him.

AUGUST 7TH. Called on [Senator] Windom and Mr. Colfax. . . .[19]

AUGUST 12TH. Reached Chicago this A.M. and received a telegram from Lizzy informing me that Mr. Roberts desired me to proceed to N.Y. without delay. Therefore I did not remain here longer than to take breakfast and left for N.Y. 9 A.M.

AUGUST 13TH. . . . [T]o New York . . .

AUGUST 14TH. . . . The [Northern Pacific] Board was delighted with my work and ordered [Bismarck to the Yellowstone construction] contracts awarded at once. I will and get off for home tonight. . . .

SUNDAY, AUGUST 17TH. Received Chicago papers of yesterday with telegraphic extracts of my report to the Board of Directors. All very complimentary. Not feeling well today, hope nothing serious. However, I ate too many peaches I suspect.

AUGUST 18TH. Reached Chicago last night feeling wretchedly. Took a warm bath and retired early. Could not sleep and I became so uneasy that I sent for a Doctor and I think it is well I did for I might have had cholera. Today I saw General S[heridan] and remained in bed for most of the day.

AUGUST 19TH. Left Chicago last night. Took sleeping car. Felt badly, telegraphed Lizzy that I will be at home this A.M. Met a Mrs. Garba [*spelling unclear*] from Virginia on train and spent a pleasant time.[20] Reached dear home 7 P.M.

---

[19] William Windom (1827–91) was a Republican senator from Minnesota in 1873. Schuyler Colfax (1827–85) was Grant's first vice president. Because of Colfax's involvement with the Crédit Mobilier, he was denied a second term. Both men were in Jay Cooke's pocket, strongly supported the Northern Pacific, and were likely pleased to see Rosser for an updated insider briefing.

[20] It is fair to speculate that Mrs. G was attractive.

Charles Loring Brace to the New York Times
THE NEW NORTH-WEST.

LAST POINT OF CIVILIZATION—DAKOTA— BISMARCK—A FRONTIER TOWN.

BISMARCK, DAKOTA, THURSDAY, JULY 17, 1873.—We are now at the last point of American civilization in the North-west, the end of the North[ern] Pacific, some 500 miles west of Lake Superior—at Bismarck, the frontier town. Beyond are the wild prairies of Dakota, the savage gorges and canyons of Montana, and the wild Sioux of the plain. The impression left on the mind on reaching here is of days on a rolling, waste ocean. The long yellow surge of the prairie surface is not unlike the long yellow sweep of the sea after a storm.

There is certainly grandeur in it, but a dreary monotony. "The Lord deliver me from ever living on it," is the first thought. The dreariness is heightened by the white bones of buffalo, scattered over the brown grass. In a few places we see emigrants, living in wagons, with the dirtiest and most wretched children, plowing up the black soil to plant the first settlement. . . .

Piles of firewood are seen at almost every station, brought from a hundred miles away. The country near the Red River—say for fifty miles—seems good land, a deep black loam, with subsoil of clay and marl, but then begins a poorer kind of prairie, more fit for grazing. The only wood skirts the few water courses, and in some of these valleys—the upper waters of the Dakota—near a few tents called Jamestown, the soil is apparently very fertile. Here the famous "bunch-grass" begins. In other portions, we pass many alkaline lakes. The Sioux—the friendly Yanktons—are encamped at various points. They give no trouble.

In one of our cars we have some seventy carpenters from Minneapolis, going up under contract to build winter quarters at Fort Lincoln, opposite Bismarck, for the troops who will winter there. Their great amusement is cracking pistols at the game we pass—the "jackass rabbits," the prairie chickens, ducks, and curlew [a small bird]. Once we saw antelope at a distance, and [on] the other side of the Red River I saw a red deer close by the road in the bush.

I asked one of them about his wages. "We git two and a half a day, sir, and for eight hours. We shall make three and a half a day easy by overwork. But you see that ain't all. We are agoin' to locate right away.

Some place around Bismarck is going to be the biggest thing this side of Omaha. Why, lots is five hundred apiece now there, and there ain't nothin' but tents and gamblers there!"

While we were talking we were put on a side track, and a special train passed us with [General Manager] Mead and some financial friends of the road inspecting it. There were only two or three sharp-looking gentlemen outside on the Superintendent's car, but they inspired unbounded admiration with my companions. "Them's the ones that does it! Them's the high-toners!"

In regard to game, they say it is wonderful how grouse are increasing on Red River as cultivation increases. The larger game, however, now center more in the Saskatchewan region— the elk, moose and buffalo. No buffalo cross the Missouri in this portion of its course. The antelope shooting in this region is excellent. They tell me that General Custer, on the Yellowstone Expedition, is coursing them with hounds, and the force get large supplies of venison through professional hunters.

Besides our carpenters, the only travelers were an army officer and one of those quiet, reflective-looking, inoffensive men, whom Bret Harte describes, but powerfully built— a professional gambler and a ruffian. The only female traveling was a somewhat gaudily dressed woman with the gambler. Both behaved with perfect decorum.

The Red River divides the journey into two nearly equal parts in point of time. The first night is spent in Fargo and the second at Bismarck. The distance, however, is 253 miles from Duluth to Fargo, and about 200 miles from Fargo to Bismarck.

THE RAILROAD

The road is well laid out until you reach the Sheyenne River; from there to Maple River it needs ballasting, and the cars sway like a boat at sea. Once our baggage-car jumped a rail, but fortunately jumped back.

The ascent from Duluth to Bismarck is about 900 feet, of which 500 feet is the rise from the Red River to the Missouri. Few roads in the United States have cost so little. The engineer assured me that it had only cost $14,500 a mile, without the rails, and with them only $23,000.

The company has done everything on large and comprehensive style. I was much struck with their roomy lodging-houses for colonists at Duluth, Brainerd, and some other points. These will eventually be of great importance in attracting emigration. They have also built their own hotels, and these have been leased to individuals. The consequence is the traveller gets comfortable rooms and good meals in every place of importance on the road. The rolling stock is the very best. Everything is done to save labor. Cars discharge at Duluth right into the steamer or the [grain] elevator; digging machines [steam shovels] do the work of seventy-five men each in the day; dumping trucks empty the dirt for the new track. The great difficulty is fuel. The road here, 500 miles from Lake Superior, finds it cheaper to burn Ohio coal from Cleveland than its own wood. Coal, however, of good quality has been discovered by this late expedition to the Yellowstone within fifty miles of the Missouri River. This will go far to remove the difficulty. The coal found up the river seems merely lignite.

They will undoubtedly grade fifty miles beyond the [Missouri] this Autumn, and be ready in Spring to lay the rails and bring the coal. The Missouri itself will be crossed for a few years by a steam ferry, and thus the immense expense of a bridge be saved.

The reports thus far of the nature of the ground in Dakota, beyond the river and in Montana, do not show any great difficulties. If they go on as they have done, the company should reach the Upper Yellowstone in the neighborhood of the National Park, in two years or more.[21] In all probability the track there will be only fifty miles from the celebrated Geysers. Still, difficulties in the financial world or troubles with the Indians may delay the work.

This expedition of Gen. Stanley will undoubtedly inspire the hostile Indians with fear, and may lead them to accept "the iron road" through their hunting grounds. But the Sioux of the plain are the last remains of "the noble savage," and are a very brave, indomitable set of barbarians. Whether they will submit to reservations remains to be seen.

BISMARCK

The definition of a Minnesota City is two bridges and a hotel. This city, however, is made up of eating shanties, gamblers' tents, and one-story wooden stores.

The title of the land is uncertain, owing to the treaty with the Indians. The future location of the bridge over the river is uncertain, and yet, such is the belief that this must be a great center of business that land already sells at the rate of $8,000 an acre among these tents and huts. The whole

---

[21]They did not reach Yellowstone Park until August 1883.

settlement is only a few months old, yet there are 147 different structures, either of canvas or wood, at least so the *Bismarck Tribune* informs us in its first number, just issued. From this journal we learn that there is a Young Men's Christian Association here, a "select school," that on "Tuesday last ground was broken for the Congregational Church, a building 30 by 40 feet; Friday it was finished, including seating; on Sunday service was held, and on Monday a school was opened!" This certainly is rapid civilizing. A steam-boat also is just in from Fort Benton, 1,300 miles up the river, having made the trip in four days, with an immense quantity of furs. Eighteen hundred miles below, and 1,300 miles above the river connections of this little place['s] reach!

The climate of the Upper Missouri fully bears out all that we know of the northern inclination of the isothermal lines as we go inland. Thus, snow was gone here in March, while it held on till May further east. By the 11th May the prairies were green, and stock was feeding on them. One old stock-breeder in the neighborhood keeps his cattle out all winter, feeding on the "bunch grass," which grows under the snow. The river was open by the middle of April.

There are no farms here yet, but the railroad company is planting a garden to test the soil. The garden at Fort Wadsworth produced wonderfully this year.

I spent the night at what is capped the "Capitol Hotel," a shanty which is both "dear and nasty." The gambling and drinking places seemed very quiet. The only disturbance during the night was from two cannon shots, the cause of which I did not ascertain, and the reports of a pistol from an Indian scout, I was told, pursuing a white soldier who had stolen his gun. It was characteristic that no attempt was made to arrest either.

Source: New York Times, *August 5, 1873.*

Pompey's Pillar, named by William Clark in 1806. Clark's name can still be found where he carved it into the rock. The Yellowstone River is out of sight but directly behind the stone monolith. Congress made the fifty-one-acre site a national monument in 2001. *Author's photograph.*

In 1872 the surveyor and budding artist Charles "Shorty" Graham (1852–1911) painted this watercolor near the Little Missouri River, one of only three known color illustrations to survive the five 1871–73 Yellowstone surveys. *Courtesy Albert H. Small Special Collections Library, University of Virginia, Charlottesville.*

Because of the Badlands, the picturesque Little Missouri River (shown here five miles south of present-day Medora) was considered a potential barrier in 1871. In quickly finding two streams that fed into the river along which track could be constructed, Rosser's surveying efforts were considered a success. *Author's photographs.*

Custer and the 1873 Yellowstone Survey, by John K. Ralston (1896–87), ca. 1970. The background of the painting was likely painted somewhere near Glendive, but on the west bank of the Yellowstone River. *Courtesy of Stockman Bank of Montana, Billings. Photograph by Larry Mayer.*

The scene of the August 4, 1873, fighting. This photograph, taken from the south side of Big Hill, shows about where the old buffalo trail ran. Honsinger and Baliran, minutes before being killed, would have ridden down to the hill's base and made a sharp left. Two miles away, in the center of the photograph, is where Custer made his defensive stand, his back to the Yellowstone River. *Author's photograph.*

Big Hill from the south, showing the possible location—the break and valley in Big Hill, all the way to the right—where Baliran and Honsinger were ambushed by Rain-in-the-Face and his half-dozen Hunkpapa on August 4, 1873. The surveyors and their two-company escort were farther to the right, where Big Hill meets the Yellowstone. *Author's photograph.*

Approximately where Custer's wagon train was placed on August 11, 1873. To the immediate right and just out of sight is the hill where Braden brought up his line of skirmishers. *Author's photograph.*

Location of Custer's charge, August 11, 1873. The line of woods to the left is where Brandon was wounded. Shortly afterward Custer lined up six companies of cavalry stretching from the trees diagonally to the left. The Sioux and Cheyennes, nominally under Gall, were about in the position from which the photograph was taken, diagonally and to the right. Just beyond where the road dips and in back of the hill line is the Yellowstone. Custer charged from left to right. *Author's photograph.*

PART III

Fighting on the Yellowstone

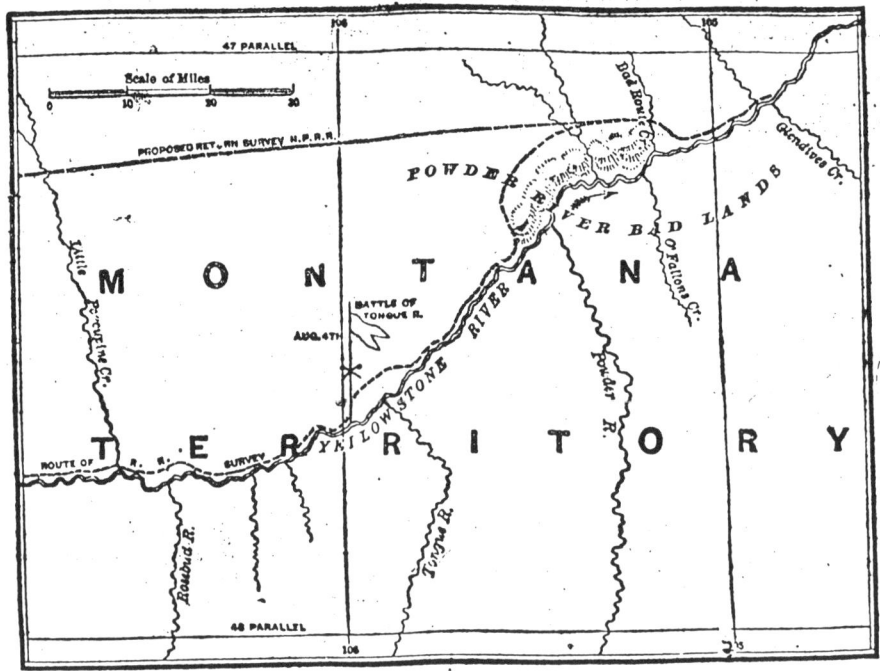

"Scene of the Battle of Tongue,"
New York Daily Tribune, September 8, 1873

This is the map that Montgomery Meigs drew for Samuel Barrows's *Tribune* article. Note how far Stanley and Custer had to take the wagon train away from the Yellowstone after leaving the Glendive Creek area. More important, readers were misled by the word "Battle," for fewer than three hundred cavalrymen and Indian warriors were involved in the failed August 4 ambush of Custer. *Author's collection.*

CHAPTER 7

Custer Ambushed, August 4

On Monday, August 4, Custer's advance cavalry force (91 officers and men) was attacked by at least 150 Sioux and Cheyenne. What no one realized was that Sitting Bull was hunting on the same side of the Yellowstone, never believing that Stanley could cross the river. Sitting Bull's Hunkpapa, Minneconjous, Cheyenne, and other followers, perhaps 500 to 600 warriors in 1873, had previously taken substantive losses in engagements with a combined Crow–Nez Perce camp, Métis, and in their Mandan ambush of the surveyors. The last thing Sitting Bull wanted was a fight against Stanley's large force.

Not until late on August 3 did an astonished Sitting Bull hear from Oglala hunters that Stanley was not just on the same side of the Yellowstone but only thirty-five or so miles away. Around dawn on August 4, Sitting Bull sent a scouting party downriver to confirm the report. When they reached the crest of Big Hill—likely near today's Miles City airport—they spotted Custer's advance party and Stanley's miles-long column. Retreating, the Indian scouts split, with some staying near Custer and others galloping back to Sitting Bull. Quickly, Sitting Bull had his camp packed and began falling back upriver toward a crossing. However, he also sent some 150 warriors to block and, if possible, defeat the advanced cavalry force.

In the clash that followed that afternoon, special interest should be taken in Custer's handling of his troops, where he exhibited none of the impulsiveness for which he is frequently accused. Equally impressive is what Tom Custer's small force did—and didn't do. Here some twelve (at most fourteen) enlisted men dismounted and formed two short lines in the face of what looked like hundreds of charging Indians. Waiting on Tom Custer's orders, they let the galloping Sioux and Cheyenne come within fifty yards before firing two volleys. While they hit nothing, it brought the attack to a halt and gave Custer time to conduct an orderly retreat.

And what if these dismounted men had broken ranks? Most likely many (if not most) would have been quickly overrun and killed and Custer's life endangered. This would probably have been followed by a melee, with both sides suffering numerous casualties. However, Tom Custer's thin line of undoubtedly frightened troopers didn't break, a testament to the Custer brothers' vigorous training and leadership earlier that year.

Chapter 7 contains one locating map and four explanatory sketch maps showing *approximate* locations and movements. Archaeologists have looked for the scene of the battle, but flooding has probably shifted the Yellowstone's course, creating islands where none existed, perhaps leaving Custer's defensive site partially or totally underwater, filled in, or "scoured," the location of its shell casings and related artifacts.

Included is Barrows's *Tribune* letter, Konopicky's often unintentionally amusing description of the day, and Bradley's brief diary entries. While in camp at Pompey's Pillar two weeks later, Custer wrote his official report of the fighting and later a second account published posthumously in *Galaxy*.[1]

A final question: Who led the Sioux and Cheyenne? While the Cheyenne participants have been identified, the Lakota leaders remain unknown, recent attempts to credit Crazy Horse notwithstanding.[2]

---

[1] Custer's report is reprinted in Elizabeth B. Custer's *"Boots and Saddles," or Life in Dakota with General Custer*, and still available in paperback. Custer's "Battling the Sioux on the Yellowstone," *Galaxy Magazine*, July 1876, is available in *The Custer Reader*, ed. Paul Andrew Hutton (Lincoln: University of Nebraska Press, 2004), and in *Eyewitnesses to the Indian Wars, 1865–1890*, vol. 4, ed. Peter Cozzens (Mechanicsburg, Pa.: Stackpole Books, 2004).

[2] Stephen E. Ambrose, Kingsley M. Bray, Marie Sandoz, and James Welch have suggested that Crazy Horse was in (and therefore led) the decoy party and the Indians who attacked Custer. Others such as James Donovan, Lawrence A. Frost, Robert W. Larson, Robert M. Utley, and Jeffry D. Wert make no mention of Crazy Horse's participation. In Mike Sajna's *Crazy Horse: The Life behind the Legend* (New York: John Wiley, 2000), he goes so far as to write, "There is nothing in the primary record to even vaguely suggest such a claim." While it would be colorful to include Crazy Horse, I have found nothing to contradict Sajna.

The source I found most useful is Father Peter J. Powell's *People of the Sacred Mountain: A History of Northern Cheyenne Chiefs and Warrior Societies*. In discussing the August 4 fighting (p. 825) Powell writes that certain Oglala took part who were "from Crazy Horse's band." While Bray cites this sentence as the source for writing Crazy Horse into the fighting and taking the lead role, the only thing I read is that Powell wasn't sure; he wasn't sure "yes" and he wasn't sure "no"; rather, he wasn't sure, period.

Other points to consider: (1) Bloody Knife didn't recognize any of the decoy party and vice versa. This lessens the chance that someone as well known as Crazy Horse

Barrows to the Tribune

THE YELLOWSTONE WAR

BATTLE OF THE TONGUE RIVER.

A HOT AND THIRSTY MARCH—GEN. CUSTER
STRIKES AN INDIAN TRAIL—CUNNING SIOUX DECOY—
TWO HUNDRED AND FIFTY PAINTED WARRIORS—
THREE HOURS FIGHTING AND A CAVALRY CHARGE—
THE KILLED AND WOUNDED.

CAMP NEAR MUSSELLSHELL RIVER, MONTANA, AUG. 19.—No day since the expedition started more monotonously than the 4th of August. Our march since we left the Yellowstone has been devoid of interest except that afforded by meeting the *Josephine*. The country presented the same arid, barren features of the eastern side, save only the bottom land along the river with its growth of timber and belt of grass. For our scientists there was nothing to do but march with the soldiers; for the soldiers nothing to do but march with the train; and for the train, nothing to do but to move slowly through the valley and climb the surly bluffs which interrupted our lines. A terrible heat had oppressed us for several days and taken the grip out of the men. The Indians, seemingly, had conspired to render the monotony more oppressive. They had kept out of sight and let us alone. We were willing they should. Only a dearly-bought experience can make one realize that these men when they are not seen are most to be dreaded. We knew that we were in an Indian country. We had seen their graves and their camp sites, and a few had been seen in the distance; yet having been out 45 days from Fort Rice without suffering a single hostile act, it was hardly possible to get up any excitement on the Indian question. A consuming ennui seemed to settle on the expedition.

---

led the decoy party. (2) The Cheyenne are known to have played a key role in the day's fighting (3) Large numbers of Oglala were in south-central Nebraska where, on August 5, they attacked an encampment of Pawnee, killing a hundred ("Lo Meets Lo," *St. Paul Pioneer*, August 13, 1873; "Indian Slaughter," *Philadelphia Evening Bulletin*, August 14; "The Recent Indian Massacre," *New York Times*, August 21).

As to Crazy Horse's riding up and seeing Custer rushing to put his clothes on, as Ambrose and Welsh insinuate—a replay of a half dozen Indians galloping at high speed over broken ground under a baking noonday sun ostensibly focusing on stealing horses but being able to distinguish one moving man among dozens in shadowed woods—well, that story doesn't make the slightest *optical* sense.

A HOT AND THIRSTY MARCH.

We left camp at the usual hour—5 o'clock in the morning. The coolness of the early morning, which had excited the hope of a relief from the oppression of the preceding days, soon disappeared, and the sun rose without a single vail of cloud. I pitied the poor men and the poorer mules, compelled to march and pull through this wilderness of sage and cactus. As I rode up to the head of the infantry line, before crossing to join a squadron of cavalry on the flank, I gathered something of the sentiment of the men concerning things in general and the expedition in particular. The army is a great school for grumblers. There are more chronic grumblers in the army than anywhere else I know of. Grumbling seems to be one of the soldier's perquisites, reckoned in with his tobacco and his pay. If he succeeds in making himself miserable in a vicarious way he creates a good deal of fun at his own expense; for there are always men enough who do not grumble to laugh at those who do. The officers are frequently as bad as the men.

"Good Lord!" said an infantry captain, in return to a "good morning" from me, "if this is not the [hard]est country that ever I saw in my life. I'd have resigned quick enough if I had known what a miserable country I was coming to."

"Curse the Jay Cookery that got up this expedition," said another.[3] Even the scientists had lost hope, and, omitting expletives, mildly referred to the wilted condition of things generally. "This expedition is getting somewhat monotonous. There is nothing to collect or see. It is too warm to write, read, or think." This was the state of opinion with the thermometer at 120° in the sun, and no shade to be had at any price. A fall of 40° would have given quite a different aspect to the march, though no condition of the thermometer could reconstruct the topography of this country.

As I galloped along I overtook Capt. Hale's detachment of cavalry, which was guarding the left flank.[4] I came up with the veterinary surgeon of the cavalry, Mr. Honsinger,[5] a fine-looking, portly man, about

---

[3] The soldier's reference is to Jay Cooke, financier of the Northern Pacific Railroad.
[4] Owen Hale (1843–77) enlisted in 1861 and died in the Nez Perce campaign. His last words were "My God! Have I got to go out and be killed in such cold weather?" Dan Thrapp, *Encyclopedia of Frontier Biography* (Lincoln: University of Nebraska Press, 1991).
[5] Dr. John Honsinger (ca. 1820–73).

55 years of age, dressed in a blue coat and buckskin pantaloons, mounted on his fine blooded horse, leisurely trotting along on the cavalry trail. No man in the regiment took more care of his horse than he. It was an extra professional care—a love of the horse for his own sake, without which no man ought to be a cavalrymen, much less a veterinary surgeon. He had taken the horse at Yankton, in the Spring, from one of the cavalry troops—a gaunt-looking steed then, but under his fostering care he had grown fat and sleek. Poor man! he was soon to make a last long camp on the lonely banks of the Yellowstone. Without dirge or funeral note, he was slowly marching to his own grave. When I saw him again, a few hours later, he was a corpse. He had died a victim to his devotion to that noble horse.

INDIAN FOOT-PRINTS.

As noon approached, the heat grew more and more intense. In addition to the heated march, the bad condition of the country compelled a great deal of work with pick and shovel. Frequent detentions at [Sunday Creek] and ravines we crossed added still more to the ennui of the day. We had camped the night before on a creek tributary to the Yellowstone. We expected to make a high detour to avoid the bluffs running down to the river and afterward to camp in the valley. Not anticipating a long or unusually hot march, the men had not provided themselves with sufficient water. Being obliged to keep well away from the river, we found none on our way. To the heat and fatigue of marching were soon added the pangs of an intense thirst. The men wilted, the mules and horses drooped. The men longed for camp and longed for water, but both were hours away.

While the infantry and the cavalry near the train were longing for any deliverance from heat, thirst and monotony, Gen. Custer, with a squadron of cavalry several miles ahead, was having a warmer and much more lively experience. With a squadron numbering 90 men [technically commanded by] Capt. Moylan,[6] one troop of which was commanded by his [younger] brother, Lieut. Thomas Custer,[7] and the other by Lieut.

---

[6]Miles Moylan (1838–1909) was captured by Rosser during the Civil War, awarded the Medal of Honor in 1894 for his actions in fighting the Nez Perce at Snake Creek in 1877, and fought at Wounded Knee.

[7]Thomas W. Custer (1843–76) was George Custer's younger brother twice awarded the Medal of Honor. For a full biography, see C. F. Day's *Tom Custer: Ride to Glory* (2002).

CUSTER AT THE YELLOWSTONE, AUGUST 4, 1873
Copyright © 2013 by The University of Oklahoma Press.

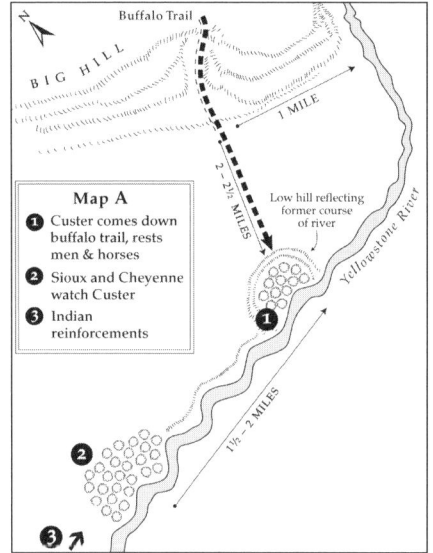
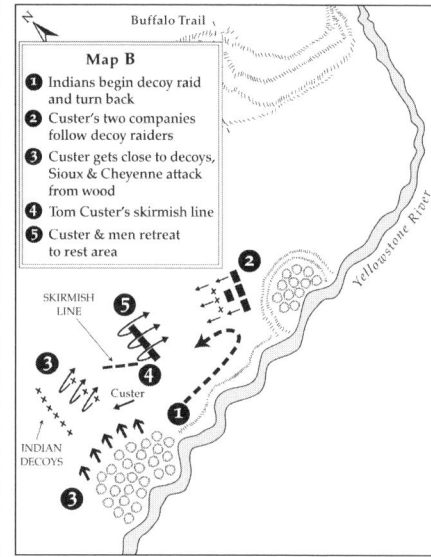

Custer arrives mid-morning and selects site for his men to rest and for the wagon train and column to camp in the evening. Becomes his defensive position in the afternoon.

Indian decoys begin "raid" on horses, Custer follows, avoids ambush and retreats.

Charles A. Varnum,[8] he had started in advance of the train to find a good road. In addition to the officers named, he was accompanied by Bloody Knife, the Indian scout, an interpreter, and the regimental adjutant, Lieut. James Calhoun.[9] When about two miles from the camp of the preceding night, they struck an Indian trail. Indians had been seen lurking near the camp several nights previous, but we had no information that they were near in any considerable force. Gen. Stanley had carefully abstained from taking the aggressive. The order was that the

---

[8] Varnum (1849–1936) was wounded at Little Big Horn, awarded a Medal of Honor for his actions in the Nez Perce campaign, fought at Wounded Knee and in Cuba, taught at West Point, and was recalled for stateside duty during World War I.

[9] Bloody Knife (ca. 1837–76) was a favorite of Custer's because of his skill and sense of humor. He was half-Arickera, half-Hunkpapa. Raised with Gall, who killed two of his brothers, Bloody Knife enjoyed killing Sioux. James Calhoun (1845–76) married Custer's sister Margaret in 1872 and died at the Little Big Horn.

Indians should not be molested unless they first molested us. Gen. Custer therefore proceeded on his march without taking note of the Indian trail, except to read the very legible caution which the pony foot-prints had left in the sand. He advanced with his detachment about ten miles at a fast walk, and, having descended from the bluffs to the valley below, halted in a pleasant cottonwood grove on the banks of the river, judging from the nature of the road he had traversed that the train would not be able to go beyond that point during the day.

AN INDIAN DECOY THROWN OUT.

It was then about 9½ o'clock in the morning. The train could not reach this point until late in the afternoon. He therefore ordered his men to water their horses, unsaddle and picket in the woods, where the fall grass afforded very good grazing. After having posted a picket on the outskirt of the wood, commanding a view of the valley and to the entrance of the wood, officers and men lay down under the trees, Gen. Custer taking the precaution to call his trumpeter and tell him to lie down within call. "I had a presentiment," said he, afterward, "that we were going to have an alarm." The camp thus secured, Gen. Custer and many of the men dropped into a light sleep. A quarter of a mile further down the river was another copse of cottonwood larger and more dense than the one occupied by the cavalry. It concealed, too, a heap of latent mischief of which the drowsy horsemen under the neighboring timber were little aware.

Two hours after, that quiet, shady rest under the grateful trees was abruptly disturbed by a sight which came to the quick eye of one of the pickets. Six Indians were coming across the plain from the woods on the river below. The corporal of the picket rushed to Gen. Custer.

"Indians are galloping this way, Sir."

"Bring in your horses, bring in your horses," yelled out the General, as he jumped to his feet.

Never was an order obeyed more promptly. Every man sprang to his picket rope, and the horses were immediately brought under cover. A line of skirmishers was at once thrown out to receive with martial courtesy the advancing Indians, whose intention when within 250 yards became clearly evident. They had not come to hold a peace conference, but to steal our horses and drive out our men. They were evidently the decoy of a larger party. Gen. Custer immediately sent word back to Capt. Moylan for all the men in the woods to saddle their horses. With his dismounted skirmish line the General kept the Indians off until the horses were saddled, the men in the timber who had saddled coming out and relieving those on the skirmish line. Gen. Custer then called for his horse, and accompanied by Lieut. Calhoun and Lieut. Custer, and his own orderly rode out toward the Indians. As they advanced, the Indians retreated. Gen. Custer, having a thoroughbred horse whose speed he had tested in many a hunt, then started in chase, taking his orderly, but telling his officers to remain where they were.

THE SIOUX LINE OF BATTLE DISPLAYED.

"Tuttle," said the General to his man,[10] "just stay behind about a hundred yards; I want to go on a little and see what those Indians are up to. But keep your eyes on those woods."

The General started. So did the Indians. They had a good start, and Gen. Custer resolved not to pursue them too far away from his men. After a sharp, short race he stopped on the plain, keeping well away from the suspicious woods. When he stopped the Indians stopped. It was evident that they would not be so audacious without a consciousness of strength somewhere. For ways that are dark and tricks that are vain the heathen Sioux is almost as peculiar as the heathen Chinese. This time the trick was indeed vain. They were fighting with no novice. As soon as Gen. Custer saw the Indian dodge, which was to use these men as a decoy to draw him into the woods, he immediately sent his orderly back to Capt. Moylan to order a platoon to dismount. Before the order could get back, 250 mounted Indians, drawn up in line of battle, came out of the woods in fine military style.[11] The 7th Cavalry could hardly have done it better. With painted faces, heads decorated with ribbons and fillets, they sallied out with loud war-whoops. Gen. Custer, putting more confidence in the feet of his thoroughbred than the voice of his rifle against 250 Indians, turned back to his command, calling out to his brother to throw out a dismounted line.[12]

Lieut. Custer had anticipated the order, and was already dismounting his men. They ran forward and took places in the grass. The Indians

---

[10]Pvt. John H. Tuttle was Custer's aide (d. 1873).

[11]Robert M. Utley, *The Lance and the Shield: The Life and Times of Sitting Bull* (New York: Henry Holt, 1993), places the Sioux and Cheyenne at "about one hundred warriors" (112).

[12]Tom Custer commanded between eighteen and twenty men.

CUSTER AMBUSHED, AUGUST 4

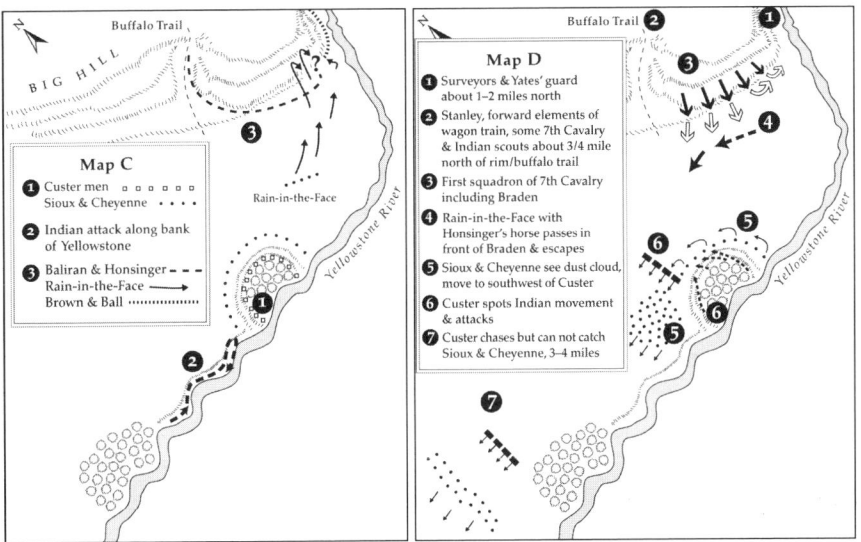

Ambush of Custer, August 4, 1873
Copyright © 2013 by The University of Oklahoma Press.

Fighting for three hours, failed Indian raid, Honsinger and Baliran route and ambush site.   *Rescue force's route, Rain-in-the-Face's retreat, Indian's fall back and Custer's charge.*

Maps by Vicki Trego Hill courtesy Research Review, Journal of the Little Big Horn Associates

opened a heavy fire, which was quickly answered by our men with the Sharpe's carbines. In a dismounted cavalry fight, every fourth man is usually detailed to hold the horses; but being short of fighting men, and the reserves being several miles back with the train, Gen. Custer ordered every sixth man only to hold the horses, and the rest to join the skirmish line. The Indians, having three times as large a force and seeing the cavalry dismounted, followed their example and dismounted. From their advantage of number, they were able to extend their skirmish line clear around from river to river, so as to enclose the cavalry in a semi-circle with the woods and the river at their back. Finding that the horses were exposed to fire, Gen. Custer ordered them to be led back further into the timber.

THREE HOURS' FIGHTING.

As he fell back with his skirmish line, the Indians advanced theirs, coming within easy rifle range. One of their number attracted a great deal of attention. He carried a long staff or lance with a pennant at the end. During the fire one of the Indians was wounded. Several others immediately collected around him, the man with the pennant, who most probably was the medicine-man, being one of the number. Lieut. Varnum called out to his men to fire into the party. The men fired, and the man with the lance fell. Previous to this, Bloody Knife, the guide, had earned the honor of the "first blood," having dropped a man from his saddle at the second fire. Gen. Custer, with his Remington, brought down a pony. After fighting the Indians in this way for nearly three hours, our ammunition gave out. Gen. Custer sent back for the ammunition of the men holding horses. While the fight was going on, the bluffs on which the train was moving with the main body of cavalry and infantry were behind the Indians. Gen. Custer noticed a party of four or five Indians moving toward the bluffs. The train had not yet come into sight, having had to make a detour to head a ravine. He was at a loss at first to divine their purpose. Presently two men were seen to run up the bluff, hotly pursued by the Indians. They were stragglers from the train.[13] He was too far away to render assistance, nor could he pursue the Indians without dividing the command and dangerously weakening it. The sad work of this small detachment was not learned until after the train had arrived.

The strategy and cunning of the Indians were not confined to extending this skirmish line and making a diversion to cut off stragglers from the train. The cavalry horses were a great temptation. They laid an ingenious plot to capture them. Having engaged the attention of Gen. Custer's whole force in front of the woods, they had sent a party down under the river bank to come into the rear of the woods and seize or stampede the horses. The first intimation Gen. Custer got of this scheme was the sight of a single Indian stealthily advancing in the rear. A few well directed shots toward him published his discovery and checked his design. He retired very reluctantly, but somewhat quicker than he came. The boldness of the single Indian was at first a matter of surprise. It was easily explained after the fight, when an examination of the river

---

[13] Pvts. John H. Ball and M. Brown left Capt. Yates's surveyor escort by moving *ahead* of the surveyors and were not part of Stanley's train.

bank showed the footprints of the fifty or more Indians which it had concealed.[14]

The Indians, having lost two men, were more cautious in their advances, and, finding that they could not with their heavy rifles drive the cavalry into the woods, had recourse to another favorite weapon. The fired the grass in four or five places. Fortunately there was little or no wind, and the grass was too short and too green to burn well, else this new weapon might have proved formidable indeed. The fire, however, raised a blue curtain of smoke, forming a corner segment between the fighting arcs. Failing in their attempt to raise a great fire, the redskins used this smoke line as a mask for their rifles. Advancing under cover of this curtain, they would pour a volley at our line and retreat. Our men soon discovered the dodge, and laid equal claim to the curtain. The Indians, abandoning this position, began to draw in their men.

CUSTER'S CHARGE.

"Now," said Gen. Custer to Capt. Moylan, "let us mount and drive them off." The men immediately mounted and advanced as skirmishers on a trot. Finding this was not fast enough, a charge was ordered. The men, eager for the order, gave a loud yell and put their horses into a full gallop. Though nearly 300 in number, the sight of 90 cavalry men coming toward them like madcaps was too much for the Indians.[15] They turned like sheep and scattered in every direction. Mounted on their fleet war ponies, with several hundred yards start, over ground with which they were familiar, they succeeded, after a close chase of three miles by our cavalry, in getting beyond pursuit. Only one of our men, Private J. R. Crow, was wounded, and he received but a slight flesh wound in the arm. One of our horses was struck in three places, none of the wounds being fatal.

Just about the time that Gen. Custer ordered the charge Gen. Stanley and those at the head of the train,[16] which was then within three-quarters of a mile from the edge of the bluff, discovered a hatless horseman

---

[14]This number appears high.

[15]In contrast, in *Galaxy Magazine*, Custer wrote: "The reader can imagine how longingly and anxiously both officers and men constantly turned their eyes to the high ridge of hills, distant nearly two miles, over which we knew we would catch the first glimpse of approaching succor [Stanley]. Our enemies seemed equally aware of our hopes and fears, and strange to say, their quick eyes . . . enabled them to detect the coming of our friends sooner than did we whose safety depended upon it."

[16]Barrows reverses the order of events.

coming full speed toward the train. His hair was streaming in the wind. His horse, without bridle, seemed moved by the same fear which impelled his frightened rider. He soon reached the train, and, almost breathless from haste and fear, told his story. He had been out hunting with three or four men of Capt. Yates's squadron of cavalry, which that day had been detailed as escort to the engineers.[17] Becoming tired, he had declined to go further with the party. He picketed his horse, took off his bridle, and let him graze, then lay down himself. Had he fallen asleep he would never have waked again. While lying here he was suddenly startled by seeing two Indians coming toward the bluff. Two stragglers from the train, one of them Dr. Honsinger, were directly in their path. He saw the Indians approach and seize their bridles, he heard a shot and saw the cavalry doctor fall. He immediately jumped on his horse without bridling him and went off at full speed. He had lost his hat, his belt, and pistol, and had come near losing his life. Gen. Stanley, who at this time had no knowledge of Gen. Custer's fight, immediately started all the cavalry in pursuit of the Indians,[18] Capt. Hale with his squadron descending into the valley at the left; Col. Hart with his squadron on the right and Capt. French[19] with his squadron on the center. Lieut. Brush of the 18th Infantry in command of the scouts was also ordered to charge down the valley. Arriving at the edge of the bluffs the curling smoke from the green plain, and the blackened patches of burnt grass showed that the cavalryman's story was not without foundation. A cloud of dust away off to the left showed the direction of Gen. Custer's charge.

TWO NON-COMBATANTS KILLED.

Two Indians of the party who had made the diversion from the main body to cut off stragglers from the train were seen making for the "Bad Lands." They were hotly pursued by the cavalry and Lieut. Brush, but under cover of the numerous ravines which afforded them a favorable opportunity of escape, succeeded in eluding their pursuers. The Indians

---

[17] Barrows didn't realize Brown was lying.

[18] According to Lt. Charles Braden, Brown was incoherent, screaming, "All down there are killed." Stanley logically interpreted this to mean that all the surveyors and Yates's escort were dead. Stanley then ordered his cavalry to rescue any survivors. Braden, "An Incident on the Yellowstone Expedition of 1873," *Journal of the United States Cavalry Association* 15 (October 1904): 228.

[19] The able Capt. Thomas French (1843–82) survived the Little Big Horn but was convicted for drunkenness (1879), leading to his army resignation.

are thoroughly acquainted with these "Bad Lands," and their ponies are used to climbing them. In such ground they are more than a match for our best American horses. Meanwhile Gen. Stanley had sent out the demoralized cavalryman [Brown] with two or three scouts to look for the bodies of the unfortunate men. The soldier's fright had scared all sense of locality out of him and stampeded the four points of the compass. He was unable to find the place he had started from. The two bodies were subsequently found by a party of infantry. They proved to be the bodies of Dr. Honsinger and Mr. Baliran, the cavalry sutler. Dr. Honsinger was shot through the body by a Henry rifle ball. Mr. Baliran was killed by two arrows which entered his back, coming out through his stomach.

Ever since we left Fort Rice, in spite of the order prohibiting straggling, many of the men have been in the habit of cutting loose from the train, to hunt game or moss agate, wandering from half a mile to two miles; but Dr. Honsinger was never known to leave the column. Mr. Baliran seldom left it. Dr. H. was the last man in the expedition who would have been selected as a victim of this dangerous habit. His absence on this occasion was easily explained. The day, as I have said before, was one of intense heat. The Doctor's horse had had no water since morning. The river lay only a mile away. The train would eventually camp on the river bottom. Gen. Custer, he knew, had gone on with his soldiers. A couple of miles back along the river was the surveying party with a cavalry guard. There seemed to be little danger in going to the river. He took the chance, risking his life to give his horse a drink of water. The Indians won the wager.

Mr. Baliran was in company with Dr. Honsinger and going to the river for the same reason. One of our Indian scouts who could not speak English met the two as they were descending into the valley. The quick eye of the scout had seen the Indians. He stopped the Doctor, and said, "Indians, Indians." "No, no," said the Doctor; "they are cavalry, cavalry." The scout took hold of their bridle-reins and tried to turn their horses back. They refused to be convinced and the scout left them to their fate. It must have been soon after that the Indians were upon them. The nature of the wounds showed that both men had been shot in the saddle. They had turned after seeing the Indians and tried to climb the bluffs. The Indians had fired and brought them both down. The Doctor, having the better horse, had nearly reached the top of the bluff when the ball overtook him. Strange to say, neither of the party was scalped. The Indians took Mr. Baliran's money, amounting to about $100, and the doctor's watch. The bodies were not mutilated. Perhaps the sudden departure of the frightened cavalryman for assistance was seen and their departure thereby hastened. Dr. Honsinger had served through the war and had previously accompanied Gen. Frémont on some of his expeditions.[20] He had been a long time connected with the 7th Cavalry and was greatly esteemed by officers and men for his personal and professional qualities. He resided in Adrian, Mich. Mr. Baliran was also a favorite in the cavalry. He has a wife and child in Memphis, Tenn.

LOSSES ON BOTH SIDES.

A third victim to the Indian surprise was Private John Ball, Company F, 7th Cavalry. He had been out with the same hunting party from which his comrade had listlessly seceded. He was never afterward seen and was probably picked up somewhere by the Indians.[21]

Our total losses in this fight, which occurred on the Yellowstone, six or eight miles above the mouth of Tongue River and takes its name from the latter stream, were three killed and one man slightly wounded. The Indians captured three horses. The losses to the Indians it is difficult to tell. No foe is more careful to conceal his losses than the Indian. One of their number was certainly killed and one was wounded. The man who was killed was either their Medicine Man or one of their chiefs.[22] As soon as he was shot they took him back to the rear and set up a fearful howling, showing, according to Bloody Knife, that they had lost one of their principal men. The next day we found an Indian shirt covered with blood, and pierced front and back by a bullet. No man who wore that shirt when the bullet pierced it could live. Either it belongs to the Medicine Man, or else another of the Indians accompanied him to the Happy Hunting Grounds.

Gen. Custer camped that night in the woods he had originally occupied and fought so long to hold. The train found passage down to the plain, and made camp about 4 o'clock. The 4th of August will long be remembered by all who accompanied the Yellowstone Expedition as one of the most eventful days of the trip. Such an excruciatingly hot day I have scarcely ever experienced. More men were prostrated by heat than on any day since we started. The death of the two men cast a gloom over this camp.

---

[20]John C. Frémont (1813–1890), western explorer in the 1840s and first Republican presidential candidate (1856).

[21]Ball's remains were discovered in September near the buffalo trail.

[22]Father Peter J. Powell, *People of the Sacred Mountain*, identifies him as a Cheyenne chieftain.

To the stragglers it was a fearful monition. "I shall issue no further order about straggling," said Gen. Stanley, "for none will be necessary."

The Indians who made this attack are supposed to be mainly Unepapas, one of the ten tribes of the Sioux nation. Of these tribes the Unepapas and the Ogalallas are essentially hostile.[23] The Santee, Yankton, Upper and Lower Yantonais, and the Brule of the Missouri are mainly friendly; also the Minnecongo, Sansaic, and Blackfeet. But in each of these tribes, except the Yankton, there are hostile remnants that gladly unite themselves with the Unepapas and Ogalallas. The Unepapas number about 2,500.

Published September 8, 1873

Konopicky to His Parents

Montana Territory on the Yellowstone
[Not dated, about September 9]

... On top of that was the terrible dust, the horrible dry heat and, most of all, the unfriendly Indian dogs who followed us almost the whole time we were on the Yellowstone.

The first time they appeared was on the morning of [August] 4. Dr. Nettre and I had ridden ahead and found us on a high prairie near the Yellowstone. We were intending to ride down into the Yellowstone Valley through a pass in the hills when we noticed that the column was not following our route but retreating to the prairie. This caused us to turn around and chase after the column. Later, we rode to a similar pass with General Stanley's adjutant and four of our Indian scouts. I was tired of riding around and fatigued by the terrible heat, so I gave my horse to a soldier and got into our wagon. I was in the wagon half an hour when the adjutant came riding down from a hill toward the column in full gallop shouting, "Turn back! The Indians are there!"

There was terrible confusion. The first wagons, including mine, turned around and raced into the herd of cattle right behind us. Even though I couldn't see any Indians from the wagon, I looked for my horse but I couldn't spot the soldier who had it. After about half an hour, the column sorted itself out. When we came to the high hill from which you could see

---

[23] The present-day spellings are Hunkpapa and Oglala.

the Yellowstone Valley, a scary sight met our eyes. The Indians had set fire to a large part of the Valley, and certainly [not] only to rob our animals of their food. This vengeful act of the Indians did not succeed because someone [unclear]. Far behind the burning meadows we could see the clouds of dust raise[d] by the fleeing Indians pursued by cavalry General Custer. We went down into the valley to make camp. Just at the beginning of the valley, I saw a group of soldiers standing around. I went over to see what was happening and found the Indians' first victim, Kaufman, who rode with the cavalry.[24] He had two arrows in the middle of his chest, which stuck out three inches in the back, and two gunshot wounds also in the chest. His dead horse was two hundred steps away at the edge of a hill [text missing][25] . . . the cavalry, both were too far ahead of the column and so became victims of the Indians, who naturally [text again missing].

Bradley's Diary, August 4–10

AUGUST 4TH. Marched at 6 A.M. Rear guard. Crossed Sunday Creek 4 times in the morning. Halted about 9 in a fine grove of trees and had a good rest, a rare chance in this country. Moved on about 10. Day intensely hot and no water after the middle of the forenoon. Men suffered terribly from heat and some of the strongest broke down. Indians appeared in front about noon and killed the veterinary surgeon and Sutler of the 7th Cavalry and one soldier who were straggling. Custer had a brush with them and drove them off killing 2 or 3. Passed mouth of Tongue River at noon, heavily timbered country all through here and the valley land pretty good. The mouth of Tongue River is a good place for a Post.[26] Camped at 6 P.M. Distance 18 miles.

AUGUST 5TH. Marched at 6 A.M. in advance. Cloudy and cool in the valley. The Yellowstone improves as we go up; fine timber on the river and hills, soil and grass good. Camped on the river at 1:30 P.M. Buried the men killed yesterday at this camp [sic]. Distance 11 miles.

AUGUST 6TH. Marched at 5:45 A.M. Rear guard. Cloudy morning, a sprinkle of rain about 9 and then the sun came out as hot as ever. Found

---

[24]Konopicky means the Seventh Cavalry sutler, Augustus Baliran.
[25]This is Murphy's First Law of Research: If something is particularly interesting or important, it will be that particular section which has disappeared. In this case, two pages (stationery size, not 8½ × 11) of Konopicky's letter are missing.
[26]This area became Miles City and Fort Keogh.

our water wagons of great service. Left the valley in the morning and went out over the bluffs. Returned to the valley in the afternoon. Made a road down a long, steep hill and had five hours hard work to get the train down. Valley very good, but narrow: fine grass and timber. Camped at 4:30 P.M. Distance 6 miles.

AUGUST 7TH. Marched at 5 A.M. in advance. Fine morning followed by a hot day. Very hot too—108 or 110 at least. Saw Indian trails all day and found the body of the fine thoroughbred captured from the Cavalry on the 4th. Passed the mouth of the Rosebud at noon and found the location of the Indian villages at the time of the fight. Camped near the Little Porcupine in a very beautiful spot; shade trees, fine grass and the pure water of the Yellowstone. Slept under an Indian grave placed in the branches of a large cottonwood tree. Camped at 3 P.M. Distance 13 miles.

AUGUST 8TH. Marched at 6 A.M. Rear guard. Crossed the Little Porcupine in the early morning, quite a stream. Had to go on the bluffs to find road for train. Returned to the valley in the afternoon and found good road. Crossed the Big Porcupine about 6 P.M. and camped on the Yellowstone. Saw some beautiful places on the river with fine timber and grass. Followed heavy Indian trail all day and found [where] their villages had been within a day or two of us. Found rifles, axes, camp-kettles and coffee, sugar and bacon scattered along the trail. Custer started after them tonight with 8 companies of the 7th Cavalry and the Indian scouts; and 7 day rations in the saddle. [B]ut as he takes beef cattle on the hoof, I don't think he'll catch Indians. Distance marched 14 miles.

AUGUST 9TH. Left camp at 5 A.M. in advance. Took to the bluffs again and reached the valley on the other side at noon, very hot day. Saw Custer with our glasses some 20 miles ahead; trails running north and west, showing that the Indians have divided. Camped on the Yellowstone at 3 P.M. near the site of old Fort Sarpy.[27] Distance 16 miles.

AUGUST 10TH. Marched at 6 A.M. rear guard. Very hot day and dusty early in the morning and fearfully hot after 10. Marched nearly all day on the bluffs and found a good road. Returned to the valley at 3 P.M. and camped about 6 miles from the mouth of the Big Horn.[28] Saw one buffalo and hope to see more soon. Distance 16 miles.

---

[27] An American Fur Company trading post in the 1850s.
[28] Near present-day Myers Bridge, southwest of Hysham.

CHAPTER 8

Sitting Bull Attacks, August 11

The deaths of Honsinger and Baliran jolted Stanley as nothing else. Not realizing how close Sitting Bull's encampment had been, and with his infantry exhausted by their August 4 efforts, the next day (Tuesday) Stanley permitted a relatively short march, immeasurably aided by cooler weather. However, once Sitting Bull's hastily abandoned campsite was discovered Wednesday and a second abandoned camp Friday, it was Stanley who began pushing Custer.

Friday evening Stanley gave Custer *all* his mounted forces—eight companies of cavalry, Lt. Brush's Indian scouts, and some headquarters staff (including surgeons)—to pursue Sitting Bull. In all there were some 460 men, including, at Custer's request, Sam Barrows, the *Tribune* reporter and Harvard Divinity School student. Then, just before starting out, Custer made a final request: he wanted to take along a few dozen "beeves" (cattle to be slaughtered), in effect guaranteeing his pace wouldn't be faster than a cow's walk. As Bradley noted in his diary, Custer would never catch the Sioux. Why? Had something spooked Custer?[1]

The chase itself, under a full moon that had risen about 8 p.m.,[2] began

---

[1] As to Custer's reasoning, the editor speculates that Custer did not wish to sack a another Indian village. Custer had to have known that his reaching the villages could result in a massacre. This had happened under his leadership at Black Kettle's village in 1868, at Baker's 1870 Marias River massacre, and at Chivington's 1864 Sand Creek massacre, in which 150 Indians were butchered. Custer knew he could barely control his troopers' bloodlust (or his own for that matter) and certainly not that of the thirty Ree scouts. Any slaughter and subsequent press coverage would have created another storm of protest and given Grant the excuse to sack him. No, I believe that in 1873 Custer pushed temptation away by literally slowing his column to a walk so that he couldn't catch up with a village filled with women, children, and old men. While Custer couldn't say anything publicly, the fact is that Stanley—a notorious second-guesser—likely agreed with Custer's decision.
[2] According to the *1873 Centaur Almanac* (New York: Crumps Label Press, 1872).

with another snafu. Since no wagons were taken, horses and mules carried food, spare ammunition, cooking equipment, and so on. The description by James Calhoun, Custer's brother-in-law, follows:

> We had not fairly started before trouble commenced with the pack horses who, not being accustomed to carrying other loads than human ones, decidedly objected to being considered mules.... The camp kettles and mess pans not being securely fastened on would bounce up and down on the horses' ribs so they must have thought they were transferred to the Band ... consequently they began to buck and in less time than it takes to tell it, their loads were scattered all over the prairie.[3]

With contents repacked, the pursuit resumed, but Stanley and Custer were unaware that Sitting Bull's camp had crossed the Yellowstone earlier in the day and was out of reach. Not until late Saturday did Custer reach the point where Sitting Bull had forded the river. Custer spent Sunday attempting to negotiate the Yellowstone, but other than a handful of individuals swimming the river, he made no progress. The reader will note that late in the afternoon Bloody Knife and other scouts built at least two "bull boats"—they looked like upside-down umbrellas—from the hides of just slaughtered beeves, but these proved too little and too late.

Perhaps this was for the better. Unbeknownst to Custer, Indian reinforcements had been arriving all day, and Sitting Bull had carefully monitored his and Stanley's progress. Had Custer been able to cross his troops in small increments, would Sitting Bull have sprung an ambush when, say, half his troops were across?

Watching Custer's troopers flounder, Sitting Bull became increasingly contemptuous. "What kind of men could these be if they can't swim a river?" he apparently asked himself. Thus Sitting Bull reached the conclusion that, since the cavalry was ineffectual and couldn't reach him, he would attack the next morning before Stanley arrived.

Beginning with Barrows's account, chapter 8 tells the story of Custer's pursuit and Sitting Bull's logical, three-pronged plan of attack that failed to take into account one thing: Custer's skills as a battlefield tactician. The chapter includes four maps showing the approximate positions and movements of Sitting Bull's forces, Custer's responses, and the role that Stanley—written out of many accounts—played in setting up Custer's

---

[3]Lawrence A. Frost, *Some Observations on the Yellowstone Expedition of 1873* (Glendale, Calif.: Arthur H. Clark, 1981), 69.

SITTING BULL ATTACKS, AUGUST 11   239

final, melodramatic charge that swept the Sioux and Cheyenne from the field.

A brief entry from surveyor A. L. Berry is included. Berry was still living in 1906 when the surveyors held a reunion in Minneapolis. Before the reunion's end, the surveyors agreed to publish a book of their collective "remembrances." While nothing came of the publishing effort, a copy of Berry's account survives. If Berry's account is riddled with errors that occur after a third of a century, its value lies in its being the only source describing why it took Stanley three hours to cover the seven or so miles between his camp and Custer's.

Barrows to the Tribune

THE YELLOWSTONE WAR.

BATTLE OF THE BIG HORN.

BURIAL OF HONSINGER AND BALIRAN—IN PURSUIT OF THE INDIANS—UNSUCCESSFUL ATTEMPTS TO CROSS THE YELLOWSTONE—THE INDIAN ATTACK—CUSTER'S ORDERLY KILLED—THE SIOUX SWIM THE RIVER AND ARE ROUTED BY THE CAVALRY.

CAMP NEAR MUSSELLSHELL RIVER, MONTANA, AUG. 19.—For three days after the fight at Tongue River, Gen. Custer with two squadrons of cavalry went on each morning in advance of the train for the double object of finding a road and looking for the Indians. It was not deemed best to bury Dr. Honsinger and Mr. Baliran in the camp where the fight occurred. The Indians have wonderful skill in finding a new grave and [then they] mutilate the dead. The bodies were therefore taken on 15 miles further to the camp of the next day, and buried that evening in the center of a square formed by four cottonwood trees. The horses were picketed over them at night, and the ground so well trampled that in the morning no trace of fresh excavation could be found. Two Henry [repeating] rifles and a pistol found on the plain by our men showed the nature of the arms and ammunition which the Indians used, and also the haste with which they had departed. The discovery of these arms had also a further significance which was still more intensified by subsequent experience. Two or three Indians might incidentally come into possession of Henry rifles through capture or theft; but two or three hundred could not be armed with these and other rifles without the guilty connivance of some rascally trader. The trail of the Indians was so fresh and well marked that we had no trouble in following it. The party that attacked Gen. Custer was entirely a war party traveling in light marching order. There was no evidence that any of their villages were near. Had we known one day after the fight what we knew three days after, the cavalry would immediately have started in pursuit of them. But on the heels of a war party with fresh horses, unencumbered by lodges or travois,[4] our cavalry would have stood but little chance of making a successful pursuit.

---

[4] A *travois* consisted of two lodge poles joining at the horse's neck and used to carry a family's possessions.

IN PURSUIT OF THE INDIANS.

Within two days after the fight two wounded ponies were found on the line of march which had been abandoned by the Indians. One of our scouts also found an Indian saddle. On the morning of the 7th Dr. Honsinger's fine bay horse was found dead on the plain, his body half eaten by wolves. The supposition that having been over pampered by his former master he had taken sick and died from the hard treatment received from the Indians was subsequently confirmed. About noon of the same day Gen. Custer, who, as usual, had gone on several hours in advance of the train,[5] came upon the site of a recent Indian village, which had evidently been a large one, numbering, according to Bloody Knife's estimate, between 400 and 500 lodges.[6] It was directly on the trail of the retreating war party.

Leading from the village the trail was three or four days old, showing that the Indians had moved their families either just before or immediately after the fight. The supposition of Bloody Knife, who reads dates and numbers, histories and prophecies, in the travois tracks, the pony foot-prints, and the whole alphabet of Indian signs and evidences as easily almost as your readers will read this letter, was that the Indians had begun to move on the day of the fight, warned by signal fires near and beyond the battle scene. The marks of such fires we had discovered on the 5th inst. It was only when we got to this village that we could understand their import. The amount of abandoned property which we picked up at their former camp showed that the Indians must have been more than usually precipitate in their flight, and that they had moved as lightly laden as possible. Large numbers of long, heavy, carefully finished lodge poles, which must have occasioned Madame Squaw many a reluctant sigh to leave behind, were found scattered over the ground. Hatchets and a large number of axes kept company with these retired lodge poles. Several paddles, some moccasins, a quantity of war paint, some Indian trinkets and playthings formed the larger part of the rest of our spoil. A quantity of coffee and sugar which some squaw in the hasty exodus had overlooked was left to the action of the elements. The cottonwood trees had been well barked by Indian ponies, and the grass nibbled very short, showing that our Indian brethren had made a considerable sojourn at the camp which had been so suddenly abandoned.

---

[5]Custer had often been fifteen miles ahead of Stanley. After August 4, his advance force was increased from two to four companies and stayed within five miles of Stanley's column.
[6]Generally there were at least two warriors per lodge.

The trail from the Indian village confirmed our estimate of the numbers it had sheltered. The tall grass for 40 or 50 yards in width had been beaten down by hoof and travois. As many as 90 distinct parallel travois trails could be counted.

AN ABANDONED VILLAGE.

We made our camp that night in the vicinity of this Indian village. The next day, Gen. Custer following this heavy trail, came upon a second village site from which the inhabitants had decamped as hastily as those of village Number One. The two parties had joined forces and proceeded together in their flight. In addition to the lodge poles and paddles there were here the framework of a medicine lodge and two or three stone mortars and pestles. The number of axes found was here increased to nearly 200. Why so many axes had been left behind would have been a problem if no such thing as an Indian agency existed. The Indian throws away what can be most easily replaced. It will only be necessary to put on a meek face and go to the nearest agency to supply the loss which the rapid advance of the United States army had occasioned.

The discovery of these village sites created a material change in our plans. There is only one way in which a hostile Indian tribe can be thoroughly crippled and intimidated, and that is by attacking and destroying one of their villages. Severe as this may seem, it is the shortest and most merciful way in the end. On discovering the design of the Indians, which was evidently to get their villages in a safe position and then attack us, Gen. Stanley determined to make an effort to punish them for their murder of Honsinger and Baliran and their unprovoked attack on Gen. Custer. Gen. Custer was confident that, if he could overtake the Indians, he would make short work of their village and repeat [his 1868] victory of Washita. Though the trail was two or three days old, there was a strong probability that by rapid marches the Indians could be overtaken, or at least could be so sorely pressed as to be obliged to abandon most of their property. Gen. Custer's report of the situation was embodied in a short note and sent back by an orderly to Gen. Stanley, who, five or six miles back, was coming up with the train. Gen. Custer's proposition to march that night was cordially seconded by the commander. "March to-night with four days' rations," was the reply. The order was afterward increased to seven days' rations. The whole force of cavalry, which, subtracting the squadron left at the stockade in the Yellowstone, numbers 459 men, was detailed for this movement.

FORCED MARCHES.

In an hour or more the train had come up and gone into camp. The scene that evening of the cavalry camp was active indeed. By 10 o'clock everything is ready. The trumpet sounds to horse. The troop commanders mount their men; then the "advance," and we march from the glowing woods into the silver moonlight in the beautiful valley before us. Three of the headquarter mess, which is composed of Gen. Stanley and staff and your correspondent, were represented on this offshoot of the parent expedition—Lieut. H. H. Ketchum, A. A. Adjutant-General; Lieut. James Jones, the General's Aid, and the writer. Our three "casuals," viz., the Hon. Louis Clifford, son of Lord Clifford; Louis Molesworth, son of Sir Paul Molesworth, and Mr. R. G. Frost of St. Louis, a former friend of the first two of the London University, who having provided their own mules wagons and outfit, were allowed to accompany the expedition, also formed part of the cavalry advance. Uniting our supplies, consisting of bacon, sugar, coffee, and hard bread, with those of the "casuals," we packed them on a mule provided for the trip by the Quartermaster. By 2 o'clock in the morning it was so cold that many had to put on their overcoats, and others had to walk to keep warm. Only once during the night was there doubt about the trail. The scouts had lost it, and the column was obliged to halt. It was soon found again. The Indians had diverged to the river and watered their horses, and then returned again to the hills.

At half-past 4 o'clock, just as daylight had fairly burnished the day, we halted in a quiet ravine for three hours' rest. We had made 25 miles. By half-past 7 we were off again, and marched until 12 o'clock, making 16 miles under a very hot sun, which brought coats and overcoats into disrepute. It was desirable to march at night as much as possible, to hide our advance from the Indians. Our horses needed grazing to make up for lack of forage. We halted here till 6 o'clock, took a sleep, had dinner, and once more resumed our march, hoping to overtake the Indians some time on the following day. After a march of six miles, when just on the verge of evening, our trail, which had followed a beautiful valley, suddenly diverged to the river. Other trails from the hillside were discovered leading down to the river at the same place. From the way in which the tall grass leaned, we could tell that the Indians from the hills had gone toward the river, instead of going from it. They had either gone to the river for water, or, what we most feared, they had converged at that point and crossed.

We looked in vain to find the trail resumed. Bloody Knife was confident they had crossed, and further, that the Indians had sent out runners to near camps and gathered in all the neighboring Indians. The question was soon decided. One of our scouts, after much difficulty, succeeded in swimming his horse across the river, and found a very heavy trail on the other side. We could go no further that night, but must lie over and attempt crossing in the morning. We went to bed disappointed. If we had started 48 hours before we did, we should have come upon the Indians when they were crossing. They had not crossed earlier than the preceding day.

THE MULE THAT FELL IN LOVE AND UNSUCCESSFUL ATTEMPTS TO CROSS THE YELLOWSTONE.

The next day was Sunday, the 10th. It was a day destined to failure and disappointment, as the succeeding day was marked for success. Breakfast as usual; then at 7 o'clock the trumpeter called "boots and saddle," and we rode down to the river. At this point, about two miles below the Big Horn River,[7] the Yellowstone is about 475 yards wide from bank to bank. Two sand bars in the river divided its waters, securing a good ford, about 200 yards wide, leaving a channel 275 yards to the other bank. It was an easy matter to wade our horses to the furthest sand bar, the water only coming up to our stirrups.

The next problem was how to get to the further shore, and the whole day was occupied with attempts at its solution.

Half a dozen men from time to time tried to swim it on their horses, but only one or two succeeded; the strength of the current and the width of the river obliging the others to turn back. Along with our party were two hunters; [Gillman] Norris[8] and ["Lonesome" Charley] Reynolds, the former mounted on a dun [dull, grayish-brown] pony, the latter on a saddle mule which he had borrowed especially for this trip. The two hunters were old friends, but their animals had been acquainted only forty-eight hours. A strong and unaccountable friendship had sprung up between them.

Now there is a false tradition in the army that the mule cannot swim

---

[7]The Bighorn (or Wind) River's flow is approximately half again as much as the Yellowstone. Why Custer made no concerted effort to find a Yellowstone crossing *above* the Big Horn's mouth is not known.

[8]Gillman "Bill" Norris was a hunter and scout with Stanley's 1872 and '73 Yellowstone surveys.

much. In view of a possible order to swim the river with our horses (an order which General Custer had too much sense to issue), Reynolds's chances of getting across with his mule were freely canvassed. Reynolds himself was not very confident of crossing, on a mule, in a swift river which had baffled the efforts of some of the best swimming horses in the regiment. "However, boys," said he, "if the old gal can make it, I reckon I can get over myself." Disgusted with the futile efforts of the cavalry at crossing, Norris, the other hunter, pulled off his boots, and mounting his Indian pony rode into the river.

This movement did not escape the notice of the mule. The thought of parting gave her unutterable pain. Reynolds, her master, was at the other end of the island. She was free to act for herself. The struggle between love and cowardice lasted only a moment; then with a sudden bound the devoted beast rushed into the river, bearing on her back, besides all her saddle equipments, three days rations of coffee, sugar, and hard tack.

A loud cheer from the soldiers, and laughter made the hills ring, greeting this new version of Ruth and Naomi.

"Hold on a minute," said a spectator, "wait till the old gal gets some water in her ears, and you'll see her turn back." But the mule had no such intention. She struck out nobly for the little dun pony, keeping her head out of water. Getting into the swift channel, the hunter's coffee and sugar soon mingled with the cold running water; the hard tack likewise, accepting a new destiny, floated down the river. Thus relieved the mule pressed on, soon overtaking her companion and swimming so near to it that Norris, fearful of getting a kick, let go his pony, and man, horse and mule raced for the shore together. It was a lively sight, and we had a lively interest in the result. If at any time the hunter and his pony disappeared in the rushing torrent, a floating and conspicuous pair of ears were never lost to sight. The fierce current carried the three far down the river; it was nearly half an hour before they emerged in safety on the other side.

I believe that Reynolds was entirely reconciled to the loss of his rations by the feeling of pride and satisfaction which this achievement legitimately created. The cavalry not able to cross, Reynolds, for the sake of exercise, swam the dangerous river and brought his mule back again with Norris and the pony. I may add that the soldiers' notion that a mule with water logged ears gets discouraged, and will not swim, is a libel.[9]

---

[9] I've borrowed this anecdote from Barrows's "The Northwestern Mule and His Driver," *Atlantic Monthly*, May 1875. The title "The Mule That Fell in Love" is mine, not the *Tribune*'s.

Men were sent out on horses to swim across, but the current proved too strong for them.[10] A small raft was made to carry across a line [rope] and thus provide a way of bringing back rafts which could be built on the other side, but the strength of the current snapped the rope several times, and the plan was finally abandoned. Meanwhile Reynolds, who with Norris, will carry this letter to Fort Benton, if the Indians do not capture them, suggested that Bloody Knife make two bull-boats.[11] Bloody Knife, by the way, is a Ree Indian from Fort Buford, distinguished for intelligence and bravery, who has been made a corporal for valuable services rendered. His limited knowledge of English compels the use of an interpreter, who is readily found in one of the half-breed [scouts].

Two beeves were immediately killed and skinned for Bloody Knife's use. Stout willow wands were then cut and the bark pulled off. They were then bent into the proper shape and bound together by thongs of raw hide. The frame, when completed, looks like a denuded umbrella frame, the tips being secured by a large willow ring which determines the diameter of the circular boat. The hide is stretched over the frame and secured to it by more thongs of raw hide. The green hide is then allowed to dry, the process in our case being hastened by a large fire. When completed, this unique boat looked like a large raw-hide umbrella, three feet deep and about five in diameter, without a handle. Yet Bloody Knife declared that this raw-hide tub would carry 1,000 pounds. A second bull-boat was made and dried in the same way. The day had been passed in these endeavors, and it was too late to make further trial that night.

We picketed our horses nearby, spread our blankets, and dropped to sleep. Just before we retired, an Indian was seen to come down to the bank on the other side and water his horse. As soon as he observed our camp he immediately fled without giving his animal a drink. His surprise showed that our presence so near the enemy had not been discovered. Had we possessed the proper appliances for crossing the river, we could no doubt have affected the passage without discovery. Once discovered, an attempt to cross a line in the morning by our bull-boats would be extremely hazardous.

---

[10] Barrows's account from the *Tribune* resumes here.

[11] The bull-boats were made from the hides of the beef cattle that Custer had taken with him.

FIGHTING BEGINS.

Whatever plan Gen. Custer might have had for the morrow, however, was subject to revision about 4 o'clock the next morning. Two or three solitary rifle shots, then a doublet, then a triplet of cracks, and finally a whole volley from the other side of the river, was the nature of the reveille that suddenly startled us from our blankets in the morning. "The Indians! the Indians! they are firing on us from the other side," shouted one of our number. We jumped up and dressed immediately. Our men seized their guns and rushed to the front, and began to fire in return. As yet only a few Indians had appeared on the opposite bank. Gen. Custer ordered our firing to stop until the Indians should come down in larger numbers and give us a better target. The horses were not moved at first, to furnish no cause of confidence to the Indians. Presently, however, a large number came down from the hills, and spreading themselves in the woods on the other side, opened a heavy fire upon our camp. Our horses were immediately moved back to the edge of the woods, and afterward to the bluff, beyond rifle range. Only one tardy man, who had tied his horse to a log and was in no hurry to move him, lost his horse. He was shot dead. A bullet struck the tree under which we had been lying, others whistled by and warned us to move our breakfast room. The Indians had attacked us just as the morning meal was ready. I had only time to seize one piece of hard tack, then rush to saddle my horse and take him out of danger.

In the mean time Gen. Custer, Reynolds, several officers, the members of our party and several of the best marksmen among the men, had taken positions behind trees and were returning effectively the Indian fire. The long range—450 to 500 yards—made accurate shooting difficult, and Gen. Custer did not wish to waste ammunition by ordering a general fire. While the firing was going on in front, Gen. Custer had stationed men on the bluffs behind us commanding a view of the river, to guard against the Indians crossing and attacking us in the rear. The firing at the river was kept up for two hours and then fell off. An incessant shooting was kept up by the Indians, and some of them held a long-range talk with our scouts while using the trees for a breastwork.

GEN. CUSTER'S ORDERLY KILLED.

"Come, man, why don't you? We'll give you all you want. We are bound to have those horses of yours anyhow. We are going to cross and take

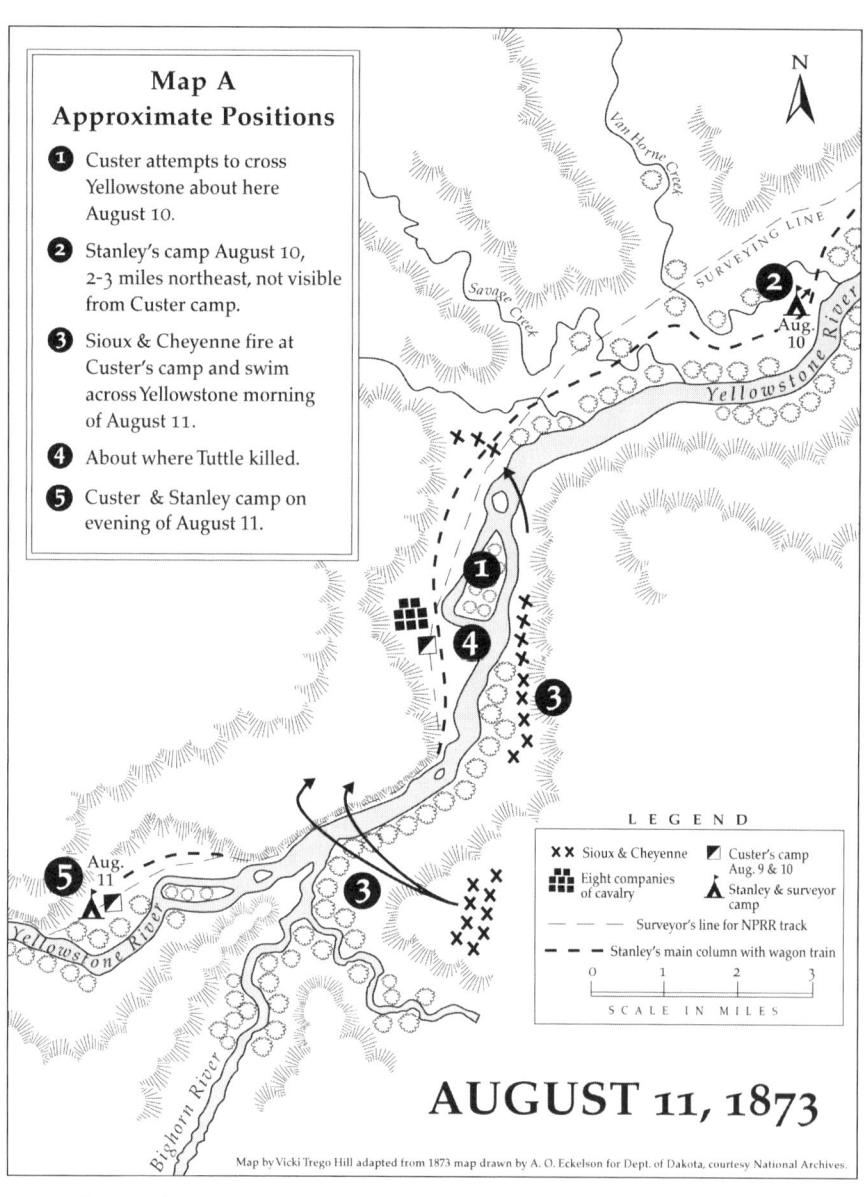

Custer's position on August 10 and morning of August 11
Copyright © 2013 by The University of Oklahoma Press.

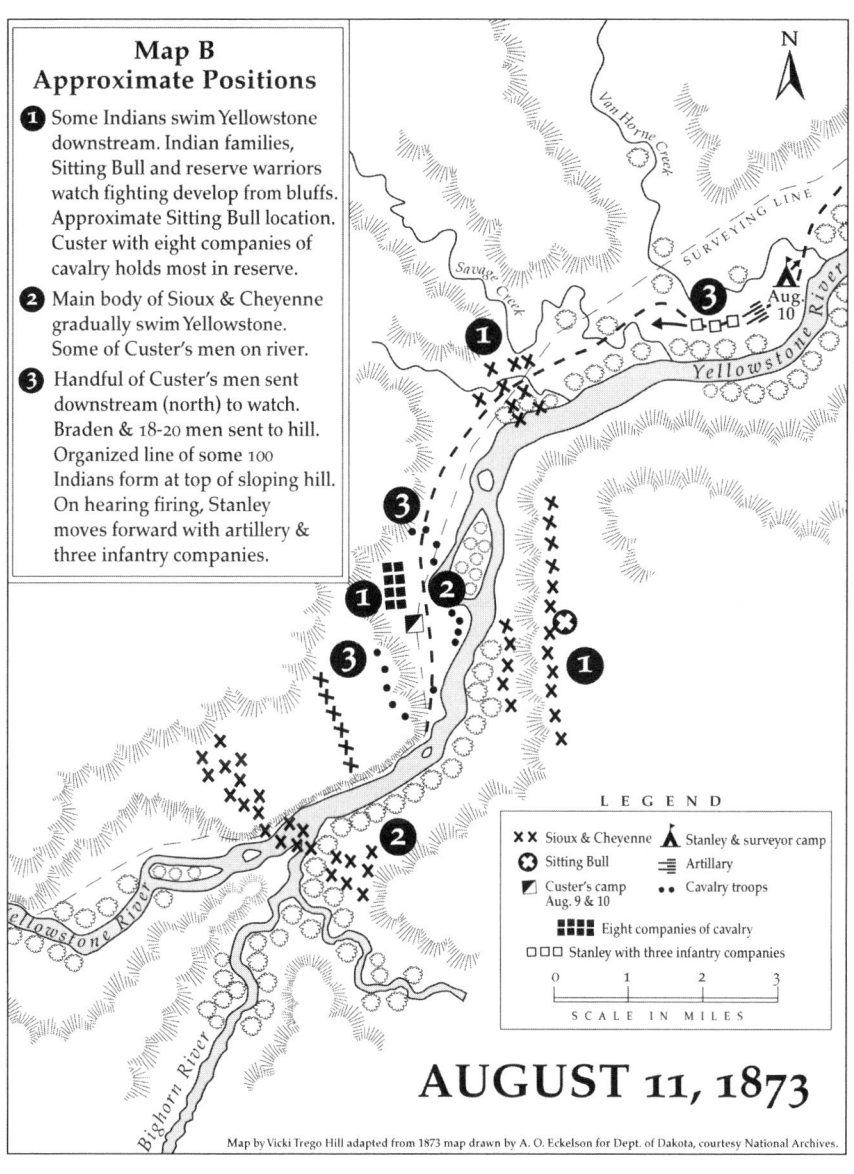

INITIAL SIOUX AND CHEYENNE ATTACKS
Copyright © 2013 by The University of Oklahoma Press.

them in spite of you," was the tone of the Indian insult when turned by our scouts into English. They were particularly desirous to know whether the two men they killed were officers. The told our scouts of the fate of Private John Ball. He had fired one shot, they said, and then had been killed. They had captured his horse, saddle, and equipments.

When firing fell off at the river, I mounted my horse and climbed one of the bluffs, where a picket was stationed, to watch the further movements of the Indians. "Do you see those black-lookin' things in the river over there, Mister," said the picket. "They are Indians."

True enough the Indians were running their horses across the river about a mile and a half below. Word was sent to Gen. Custer. The force watching the hills, those below under Capt. French and those above under Col. Hart, were immediately strengthened by reinforcements. Lieut. Brush, commanding the company of scouts, which had been ordered to proceed with the cavalry, was directed to send his men well out from the river and watch the Indians closely.

The hostiles, hoping to divert attention from the movement, opened fire again heavily on the river. Gen. Custer returned to the woods and took a few more shots. Private Tuttle, his orderly, taking up his rifle dropped down behind a tree and opened on some of the Indians who were defiantly exposing themselves on the other side. He had been firing in the morning and had got the range of the river very well. He was one of the finest shots in the regiment, and generally admired by all his comrades for his skill with his rifle. Col. Hart's orderly and another man were with him, concealed behind a large tree. "I want to drop a few more of them fellows," said he, "before I leave here. Do you want to see me drop that Indian?" pointing to one on his pony who had come out of the wood. He brought his rifle to his shoulders and fired. The Indian dropped from his pony. Three or four Indians ran to him. Tuttle took another shot and dropped another man. A short interval and another man fell from his pony. This was Tuttle's last shot. The Indians brought their best man forward and opened a cross fire on the tree behind which he was concealed. He turned his head a little on one side to take another sight, when a shot struck him in the forehead and pierced his brain.

THE INDIANS SWIM THE RIVER.

Just about this time the men who were stationed on the hills began to wave their hands, showing that the Indians were coming up the river and crossing. Gen. Custer, who was under a tree in the rear of the center, saw the sign, and immediately ordered Capt. Yates to send a platoon up to the point commanding the bluffs. Gen. Custer directed him to take Lieut. Braden.[12] The position assigned Lieut. Braden was a very important one. [Facing west, i]t was a knoll of the extreme left, that commanded a view of the front and the whole valley of the river. Capt. Yates was immediately afterward directed to dismount the remaining portions of our company, and throw them forward to hold the bluffs in his front. Gen. Custer then rode up on the bluff. As he rode up he saw the Indians approaching, and sent back word to Lieut. Weston, who commanded the other troop of Capt. Yates's squadron, and directed him to move his men up the ravine toward the bluff on which the Indians were to be seen, with directions to report to Gen. Custer soon after. He was informed that, when opportunity occurred, Gen. Custer intended him to charge on the Indians.

Meanwhile Lieut. Braden and his men were having a hot time. The Indians had gained the top of the bluff up the river, and advancing rapidly, were driving our scouts before them. The narrow ridge which Lieut. Braden occupied with his 17 men formed a natural passage across a deep ravine. The scouts made for the ridge, to avoid the ravine on one side and the steep cliff on the other, the Indians closely pursuing them. "Lie low, men," said Lieut. Braden; "wait till they come close, and then let them have it." Gen. Custer, seeing the way the Indians were coming, sent Lieut. J. H. Jones,[13] who acted as his aid, to tell Lieut. Braden to hold the place at all hazards. Braden had a picked lot of men, whom he could well trust. They heard the order and determined to hold the place.

The Indians, all unconscious of the men behind the ridge, came on at full speed. Then within 25 or 30 yards, Braden gave the order to fire. His men obeyed. Two or three Indians fell from their saddles, and the remainder broke right and left for the ravine. Our men had anticipated this movement, and kept up such a brisk fire that the Indians scattered in all directions. Rallying, however, they poured a volley into our men, one ball of which struck Lieut. Braden in the thigh, breaking the bone and inflicting a severe wound. Lieut. Braden immediately fell and rolled some way down the ravine. As he fell, some of his men ran to him. "Men,

---

[12]Lt. Charles Braden (1847–1919), West Point '69. See bibliography for his accounts.

[13]Lt. James H. Jones (ca. 1846–1919), West Point '68. Independently wealthy, he resigned from the army in 1874.

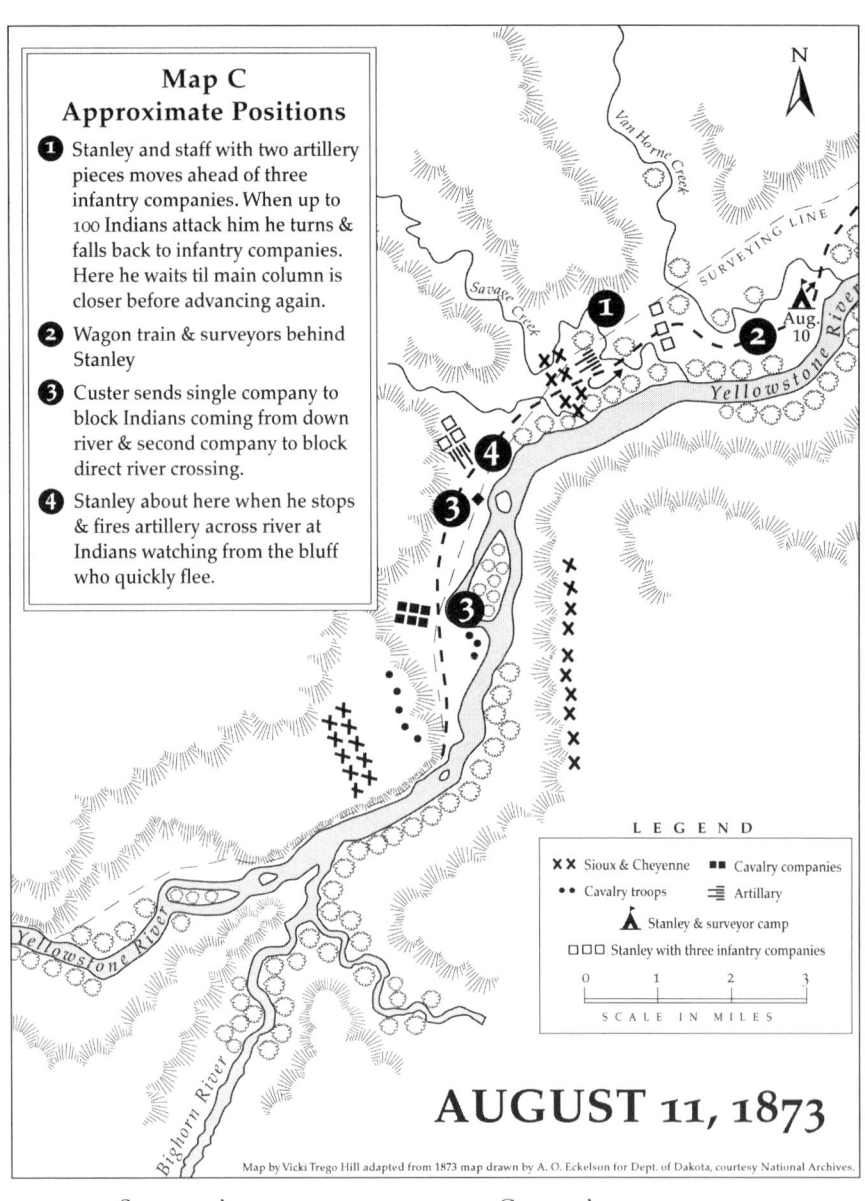

Stanley's initial arrival and Custer's first moves
Copyright © 2013 by The University of Oklahoma Press.

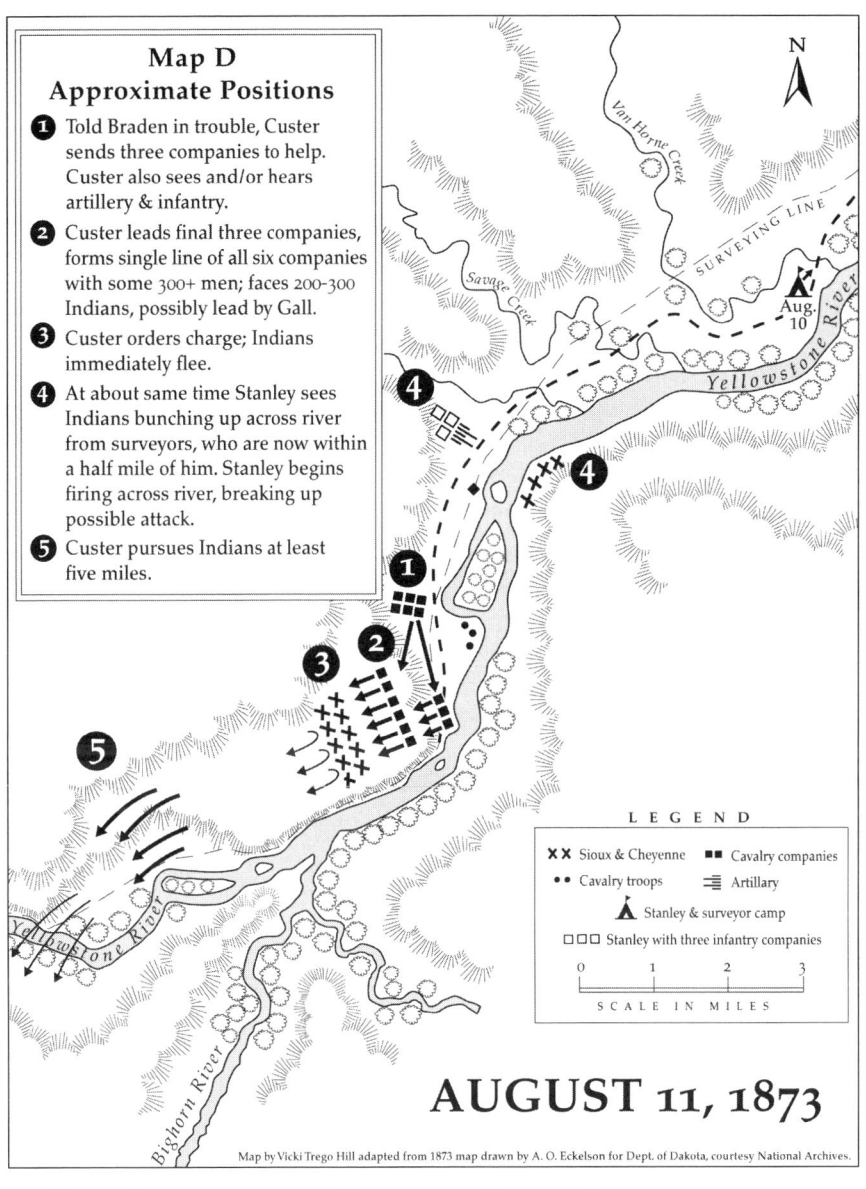

FINAL INDIAN ATTACKS; CUSTER AND STANLEY'S COUNTERATTACKS
Copyright © 2013 by The University of Oklahoma Press.

I want you to hold that ridge," he said. "You bet we will," said the men. "If we move at all, Lieutenant, we'll move forward." The men were as good as their word. The 200 Indians that had advanced on this little party were completely repulsed. Several of the Indians were killed, and some of their horses were shot.

THE CHARGE AND ROUT OF THE SIOUX.

The second onslaught of the Indians was made on our center. Failing to dislodge Lieut. Braden, a large party of them got into a ravine on our front and opened a galling fire upon us. Gen. Custer, who had established his headquarters on the top of this bluff, now deemed it time to drive the Indians away. "Strike up Garry Owen," said he to the leader of the band.[14] The familiar notes of that stirring Irish air acted like magic. If the commander had had a galvanic battery connecting with the solar plexus of every man on the field, he could hardly have electrified them more thoroughly. What matter if the coronet played a faltering note, and the alto-horn was a little husky? There was no mistaking the tune and its meaning.

"Forward!" shouted the commanders, and away they went "pell-mell," the horses seeming to share the eagerness of the men. There was no scattering or flagging. Everyman keeps in his place. On they go like a whirlwind. Weston and Hale on the left, Tom Custer plunging down the ravine on the right. Ditches, gullies, hills, cannot stop them. Now Yates and Moylan are ordered to advance and support them, and add two more to the black columns dashing over the hills. No Indian would venture to stand before that whirlwind. Leaping on their horses they run for their lives. Hale and Weston succeed in getting within thirty yards of them, and give them a volley from their carbines; McDougal speaks with his revolver.

One conspicuous Indian in a red blanket, supposed to be Gall, an important chief, had his pony shot dead under him. He leaped on a fresh horse and got away. Seeing that with their jaded horses, wearied by the long march, they could not run down the Indians, they dismounted and gave them a volley from their carbines, then remounted and chased them till they had completed a run of eight miles. Lieut. Custer on the left, led his men on furiously. "He is a terrible rider," said one of our

---

[14]The band leader, Italian-born Felix Vinatieri (1834–91), enlisted in 1861. Pure trivia: Vinatieri was the great-grandfather of Adam Vinatieri, a professional football player and Super Bowl–winning field goal kicker with the New England Patriots and the Indianapolis Colts.

"casuals" to me. "I saw him fly over a ditch about 15 feet wide. The man after him missed it, and horse and rider rolled into a gully." Two or three times the Indians ahead of him, having a better start than those fleeing before the retribution of Hale, turned in their saddles and fired a volley which tore up the dust at the feet of the pursuers, but did no further damage. They were soon too closely pressed for this amusement, and concluded to affect their escape through discretion rather than through valor. While the charge was taking place, Capt. French with a squadron had taken position further down the river to prevent Indians crossing and attacking us in the rear. A brisk fire was kept up with the Indians on the other side. A well-directed shot from French's Springfield rifle knocked an Indian off his pony. A few similar shots from his best marksmen cleared the bank on the other side in a short time, those who were not hit concluding that it was best to retire to the hills.

STANLEY'S ARRIVAL WITH ARTILLERY.

Just at this time an advent was proclaimed that greatly delighted us. We had been away from the train for three days. Our rations were pretty much gone, our ammunition considerably reduced; yet owing to the bad country we had come over, filled with creeks, ravines, and lofty hills, we did not expect the train for 24 hours. When, then, an officer came dashing up and said it was in sight, we could scarcely believe our senses. But, sure enough, the line of white bonnets two miles away was coming down into the valley. A cheer went up from the cavalry reserves. "Here comes old Standby.[15] Bully for him. Lord, what a march he must have made with that train." "Here come the dough boys.[16] They'll give the reds the devil if they get a chance." No wonder the cavalrymen were glad to see the train. It meant rations, ammunition, long guns, and artillery. There is no gun that the Indians are so much afraid of as our long Springfield breech-loading rifle with which our "dough boys," or infantry, are armed. It kills at a thousand yards.

---

[15] During the war Stanley's affectionate nickname had been "Our Stanley" or "*ünser* Stanley" by his German-speaking Wisconsin regiments. "Old Standby" was clearly the derogatory nickname given by the cavalrymen, and Barrows's inclusion of it is surprising. Or perhaps it simply reflects Barrows's disgust with Stanley's drinking, so he slipped it into his description of the fight.

[16] "Dough boys" was the World War I equivalent of GIs in World War II; this is one of its earliest uses and an idiom not usually associated with the Indian Wars.

On the other side of the river, half way up the hill, two miles away from us, was a crowd of Indians, many of them squaws and old men, who had gathered to witness the fight, and a large number of warriors who had been driven from the river.

"What a splendid artillery shot that would make," said an officer.

The idea had occurred to someone else.

"Bang! whiz!"

"Halloo! Stanley's talking Dutch to them. Look at that!"

The shell went whizzing through the air. A cloud of dust marked where it fell on the other side. When the dust cleared away the Indians were gone. Another shot to explain the first, and a third to emphasize the second, and the battle of the Big Horn was over.

[Here Barrows's account abruptly ends.]

Published September 9, 1873

Konopicky to His Parents

Montana Territory on the Yellowstone
[Not dated, about September 10]

For three days we have been back at the Yellowstone, at the same spot where we first glimpsed it. Only three more weeks and I will be back at Fort Rice from where I will be back in Cambridge by rail in eight or ten days. The biggest and hardest part of our journey is behind us and that makes me very happy. Yes, my dears, it was a much more arduous journey than I had imagined. I had no idea of the mostly desolate country that we wandered through. Maybe you can picture what I am describing, that the path the expedition took is, from Fort Rice on, almost nothing but hilly prairie, treeless, arid, and interspersed with horribly jagged sand mountains. On top of that was the terrible dust, the horrible dry heat and, most of all, the unfriendly Indian dogs who followed us almost the whole time we were on the Yellowstone. . . .

Around noon, we saw a large group of Indians sitting on a mountain on the other side of the Yellowstone. This caused General Stanley to have one of our cannons brought up to send a round of grapeshot over.[17]

---

[17]Stanley's two rifled artillery pieces were using shrapnel or similar explosive charges, not grapeshot.

It was a magnificent sight to see the shot (the distance was about 1,500 steps) land right next to the sitting group and explode. We couldn't tell whether any Indians were killed, but they dispersed in a terrible hurry.

About an hour later, we came upon the cavalry, resting after the fight. General Custer came riding toward us with his band and a flag bearer with the American flag flying. He was on a white horse and wore a red shirt. He is a clear target (he wears his hair long), but the Indians have not yet succeeded in getting him, even though he always rides in front leading his soldiers. The cavalry's losses were: one man dead, one officer seriously wounded, about six men with minor wounds and three horses dead. About 20 Indians were killed and many wounded.[18] It is peculiar that the Indians never leave their dead or wounded behind, but load them hastily on horses and take them with them. We were quite close to the place where the fight took place, on the Yellowstone. After General Custer told General Stanley that the woods were full of Indians, the latter ordered the cannon to fire shells into the copse of trees....

[Here Konopicky repeats the story of the cattle being spooked by wolves and shots being fired, which many assumed—including Konopicky—were Indians sneaking up on them.]

Next morning, we discovered what the shooting meant. They were spooking our cattle, which became skittish and stampeded. After a head count, we found no fewer than 40 of them missing.

That day we headed for the Musselshell River, which we reached in three days. This was the most interesting part of our journey, in that we began to see at first a few, but by and by, huge herds of buffaloes. We estimated that one herd had 2,000 to 3,000 head. Our Indians took the first victim from these animals, a magnificent bull that weighed about 20 hundredweight. The most important came the next day—I got one of these animals, and despite the fact that I was not armed. I had ridden a little ways ahead with Dr. Nettre and we came upon a ditch. To my surprise, a wounded buffalo was sitting upright in it. I dismounted, took Dr. Nettre's revolver, walked up to the beast and put four bullets through his head, without any effect. Only after a fifth shot behind the shoulder did the animal collapse. Too bad I did not have a knife to cut out the tongue (it is a delicacy). We are now eating much buffalo meat, but also other game, such as antelope, big horn sheep, deer, etc., etc.,—the table is full.

---

[18] Konopicky's estimates were far too high.

We followed the small but picturesque Musselshell River for eight days, during which three grizzly bears were shot. Today we are already on the right bank of the Yellowstone. We were ferried over late yesterday evening and, because of the late hour, had to sleep without tents. As I hear, tomorrow we start back to Fort Rice. . . .

Many greetings to everyone, etc., etc.

Bradley to His Wife

No. 20
Camp on the Musselshell
August 19, 1873

My darling love:

You will see from my heading that we are making progress toward the end of this long, long, weary march. We have turned our faces towards home at last and twenty days more ought to see us at the crossing of the Yellowstone again. Two of our hunters, Reynolds and Norris, leave tonight for Fort Benton[19] with dispatches and will take one light letter for each Officer. So this is the only one I can write and is the last you will get perhaps for, I hope the next word I send, will be a dispatch from Fort Buford that we are on the way home. . . .

We reached Pompey's Pillar on the 15th, rested there a day and started north for this point on the 17th. From here we march due east and try to reach the Yellowstone at the point where we crossed it, but as the country between is unexplored, there is some doubt about our getting through and in case we find it impracticable, we shall make our way down to the Yellowstone again and return on our own trail. The difficulty we expect to encounter is want of water. This interior region between the Yellowstone and the Upper Missouri is supposed to be a dry country and as we consume 40,000 [*sic*] gallons of water a day, it won't do for us to be caught in a dry country.

We reached here at 3 p.m. and are camped in plain sight of the snow peaks of the Judith Mountains, the Musselshell is a pretty stream and the water is as clear as the Cache [*two words, illegible*]. Our march from the Yellowstone here was 60 miles and we are about that distance west of the big bend of the Musselshell, the point we start for tomorrow. We

---

[19] Fort Benton was the nearest location with telegraphic services.

had some fighting with the Sioux on the Yellowstone, they attacked us twice and were driven off with loss to themselves both times. In the first attack, Aug. 4th, they killed the Veterinary Surgeon and Sutler of the 7th Cavalry. Simple murders as they were separated from the Regt. and unarmed. On the 11th, they attacked Custer near the mouth of the Big Horn. He lost 2 men killed, one officer, Lieut. Braden and one man wounded. Custer fought them 4 or 5 hours 'till we came up and drove them across the Big Horn. You may see mention of this in the papers, but it was a small affair and you must not feel any anxiety about Indians. They don't venture near our main column and always run when it comes to real fighting. Since we left the Yellowstone we haven't seen an Indian and I doubt if they show themselves again. The valley is full of buffalo and we are killing them every day.

Bradley's Diary, August 11–18

AUGUST 11TH. Marched at 5 A.M. In advance. Just after leaving camp, heard rapid firing four or five miles in advance. Knew Custer was engaged and hurried up to join him: a few Indians in sight a couple of miles off. About ½ past 8, saw about 100 in a body on the south side of the Yellowstone, threw a couple of shell over, and stamped [*sic*] them. Came up with Custer about 10 A.M. Indians hurrying off. Cavalry after them. Sent Infantry companies to command the timber and put section of rifled guns in position. Indians got across the river as soon as possible. Cavalry had one man killed, one wounded and Lieut. Braden severely wounded, thigh bone broken. Camped opposite the mouth of the Big Horn. Very hot day. Distance 10 miles.

AUGUST 12TH. Marched at 6 A.M. Rear guard. Morning cloudy and cool. Had a rough road, most of the day on the bluffs; passed the mouth of the Big Horn at 9. The river and valley looked very pretty. Camped in the valley at 5 P.M. Distance 16 miles.

AUGUST 13TH. Marched at 5 A.M. In advance. Road led over the bluffs and was pretty good until noon. Latter part of the day very rough. Could not get to the river, country impassable for teams. Camped on unknown creek at 6 P.M. A heavy storm came on just as we made camp. Custer's men struck a band of elk and killed sixteen. Engineers and escort did not reach camp tonight. Distance marched 20 miles.

AUGUST 14TH. Lay in camp all day waiting for engineers. Cloudy and cool all day. Company of cavalry sent out to find road to the Yellowstone.

AUGUST 15TH. Marched 6 A.M. Rear guard. The last march westward and everybody glad of it, for it has been a long, weary road. Came in sight of Pompey's Pillar at 9 A.M. Camped about ½ mile from the river on a little flat surrounded by high, rocky bluffs. We are disappointed in the Yellowstone country, the valley is narrow, rarely two miles in width and bordered by high bluff[s], dry and barren. Above the Tongue River the valley improves, grass and timber are plentiful, and the soil is good, but away from the valley the country is dry and sterile. Above the mouth of the Big Horn and above the hills are well covered with pine. The Yellowstone is a fine river and navigable in the season of high water as high as Pompey's Pillar, probably. The water is as pure and sweet as I ever saw. Pompey's Pillar is a rough looking butte of drab sandstone about 125 feet high[20] and 150 at the base standing on the south bank of the stream. Must be a prominent object approaching from the south. Distance marched from the crossing of the Yellowstone to Pompey's Pillar—237 miles.

AUGUST 16TH. Lay in camp all day waiting for Engineers to finish up their work at this point; connecting their line with last year's survey from the west. Indians came up on south side of the river and fired into camp in the forenoon. Stood to answer and drove in the herd. Nobody hurt. Rain in afternoon. Had an alarm at 11 at night, sharp firing on the picket line, proved to be caused by stampede of the cattle herd. Probably scared by wolves among them.

AUGUST 17TH. Marched at 5 A.M. In advance. Struck out northwest for the Musselshell. Marched all day over a high, dry divide with no water and little grass. Road pretty good. Engineers ran levels across to determine altitude. Camp at 2 P.M. on unknown creek with little water. Distance 20 miles.

AUGUST 18TH. Marched at 6 A.M. Rear guard. Fine cool morning and comfortable day. Marched over high rolling divides. About noon crossed a high rocky ridge, the highest point between the Yellowstone and Musselshell. Camped at 5 P.M. Saw plenty of buffalo in afternoon and killed a number. Distance 20 miles.

---

[20] Actually Pompey's Pillar is 200 feet tall. Bradley's meaning of "150 at the base" is unclear.

Meigs to His Parents

Camp Pompey's Pillar
August 16, 1873

My dear Father and Mother

I think my last letter was sent from Powder River and I described our adventures up to that time pretty fully. We have two newspaper reporters with the expedition and to them I refer you for a description of the battles etc. of the trip. As I have never seen any of either of their articles I cannot say how truthfully they will prove, but I think the *Tribune* reporter [Barrows] will write a pretty fair article.

The country has much improved in appearance since leaving Powder River. The bluffs along the river instead of being precipices of clay are now picturesque masses of sandstone water worn into the most fantastic shapes. The Valley of the Yellowstone until we got within ten to fifteen miles of the Big Horn has been generally five or six miles wide; below the Big Horn though, it is not more than one and a half to two miles from bluff to bluff. Indeed during the Big Horn fight while we were still two or three miles from where Custer was engaged and could hear the volleys of musketry quite distinctly. [F]ive or six Indians dashed up onto a peak on the opposite side of the river not more than three or four hundred feet off and yelled and brandished their guns in a most hostile manner. It was useless to shoot at them though we could hear distinctly what they said.

One of our scouts went down to the river bank and asked [an Indian] if Gen. Custer had killed their women and children the night before. The Indian said no, but that they were fighting Custer at that moment and that their warriors thought it was a mere pastime to get away with our cavalry. This was rather a one-sided statement as we now know of fifteen or sixteen [of them] had [been] killed or wounded and we had but three. We have seen none since the Big Horn until this morning when six galloped out on the bank of the river and yelling, fired their guns into a lot of soldiers fishing, washing and swimming on our side. There was quite a scatteration on our side, you may imagine, but the Indians were perfectly safe and scampered off behind Pompey's Pillar in an instant and before anyone could shoot back. We could see them long afterward running away across the prairie on the south side of the river.

The Yellowstone is here a lovely stream. Its water is limpid and bright and flows at the rate of six or seven miles per hour. Right opposite the mouth of the creek in which we camped is Pompey's Pillar, a detached mass of sandstone a hundred and fifty feet high with sides vertical and water worn on all sides but one. The river flows between it and the high rocky left bank in a stream some five to seven hundred feet wide. Pompey's Pillar is a very remarkable landmark and well named though it resembles more some of the ruined tombs about Rome than it does [the real] Pompey's Pillar.

We are all protected against Mr. Lo on our marches and have both cavalry and infantry strung out along our line while we work every day. We all like the cavalry officers very much. There is a dash and a captivation about the cavalry service that would make, it seems to me, every Infantry officer envious, and indeed the cavalry do despise the doughboys thoroughly, while the infantry hates them as cordially. The Seventh is a gallant regiment and Custer, albeit a little vain and fond of notoriety, is a Colonel to lead them to victory.

Tomorrow we strike for the Musselshell where we will begin our return line to Glendive's Creek. I made a small map for the *Tribune* reporter which will perhaps be published in connection with his report, and you will see our proposed line marked on it.[21] We are getting on fine in the party and have averaged on the days we worked some eleven or twelve miles[, a] great deal of the line being over very rough ground. I think the R.R. line ought to come up the Yellowstone but on the south side of the river—the north side being obstructed by steep bluffs which it will cost a great deal of money to use for R.R. purposes. Then there is any quantity of Norway pine on the hills above the river to make ties for this part of the R.R. and easy transportation for them by water to convenient points while on the Musselshell River the country is supposed[ly] very barren and difficult.

We begin to see more game and bands of elk of from seventy-five to a hundred fifty have been seen lately. Gen. Custer ran into a herd of seventy-five day before yesterday and killed thirteen, and yesterday I saw a buffalo killed not a quarter of a mile from the tent I am writing in. We expect to see buffalo on the Musselshell. . . .

A. L. Berry's "Remembrance"

. . . The next morning, early, the Indians returned and attacked Custer from both sides of the river and killed a great many of his horses. The county being so rough, he was unable [at first] to charge them on horseback and had to do all his fighting on the ground. The balance of the

---

[21]Meigs's map was printed in the *Tribune* on September 8 and is reproduced in chapter 7.

expedition heard the firing [that same] morning about [five] o'clock[22] and we hurried forward as fast as possible to give Custer the assistance which he required, as the east bank of the river seemed to be black with Indians, as well as the hills in all directions.

As soon as we came in sight of Custer's Camp, General Stanly ordered the artillery back on to the high points so as to be able to shell the woods and gulches. He sent with the artillery two or three companies of infantry to support them. The artillery started off on a wild dash, whipping the mules into a run, and the infantry followed as best they could. Before they had gone over a mile, however, the artillery were considerably ahead of the infantry, who were struggling far behind, and before we could hardly realize it, a body of Indians galloped out of the gulch right in front of the artillery, and it looked as if we were going to lose both pieces of artillery as well as the men [including Stanley]. The mules were turned around as quickly as possible and started back on a run to the infantry, who formed in line as fast as they could get together and began firing on the Indians and managed to kill some of their horses before they could get to the artillery. It was a very exciting chase, but as the artillery made quite a detour, the infantry managed to get partially between the cavalry and the Indians and cut the Indians off.

As soon as the balance of the guard could come up the artillery started ahead and soon reached a point where they could shell the woods and the opposite bank. Our Engineer Corps, under a heavy escort, had also come up and we were protected by a high bank almost under where the artillery was stationed. The artillery now began firing rapidly into the woods and other points wherever they could see the Indians moving, and as the shells would burst it seemed to terrify them, scattering them in all directions, and within an hour after we came up they were effectually driven off.

I remember distinctly the bursting of a shell which was aimed at the east bank, where a lot of Indians were congregated on a shelf rock. The shell burst on the bank about ten feet over their heads and carried down a great deal of stone, rock and dirt into their very midst. Some of them plunged off the rock into the water below, a distance of at least sixty or eighty feet, while the balance made a wild scamper to get out of the reach of our guns. . . .

Source: A. L. Berry, "The Yellowstone Expedition of 1873," Thomas L. Rosser Papers, 1171-g-h-j, box 12, Small Special Collections Library, University of Virginia, Charlottesville.

---

[22]Berry's 8 A.M. figure is incorrect and has been changed to 5 A.M. based on Bradley's diary.

Lt. Col. George A. Custer holding the antler of a slain deer. This is perhaps the only known photograph of Custer taken while on the 1873 Yellowstone Survey. *Courtesy National Archives.*

Samuel J. Barrows. Photo taken ca. 1890 when Barrows was a clergyman with the Massachusetts State Militia. *Author's collection*.

Col. David Sloane Stanley, commander of the 1872 and 1873 Eastern Yellowstone Surveying Expeditions. Stanley's commendable effort in 1872 was marred by his drinking, and his 1873 leadership remains controversial. Although William Pywell is known to have photographed Stanley during the 1873 survey, that picture is missing from the National Archives. This photograph was taken ca. 1880. *Courtesy U.S. Army Military History Institute, Carlisle, Pa.*

Northern Pacific surveyors, 1871 or 1872. Tom Rosser stands in the doorway in the light-colored suit. The group's informality and camaraderie are apparent. *Courtesy Albert H. Small Special Collections Library, University of Virginia, Charlottesville.*

Luther P. Bradley, ca. 1890. Bradley was a bookstore owner and salesman before the Civil War, but he rose to brigadier general in the Regular Army because of his ability and bravery in combat. Sheridan thought highly of him, and he was third in command in 1873. Few officers on the Yellowstone survey had seen as much combat as he. Bradley's diaries and letters offer an insightful, pro-infantry view of the survey. *Courtesy U.S. Army Military History Institute, Carlisle, Pa.*

Montgomery Meigs at the eastern Yellowstone surveyors' reunion in Minneapolis, January 1906. A major purpose of the event was to honor Rosser, who had been felled by a stroke the previous year. The reunion was well attended and given extensive coverage in Minneapolis and St. Paul newspapers. *Courtesy Albert H. Small Special Collections Library, University of Virginia, Charlottesville.*

A. L. Berry at the surveyors' reunion in 1906. Berry's years-later description of the August 11 fighting is the only known account that discusses both Stanley's and the surveyors' movements. *Courtesy Albert H. Small Special Collections Library, University of Virginia.*

W. Milnor Roberts was one of the era's leading civil engineers, and his 1869 reconnaissance and report gave Jay Cooke the final proof of the Northern Pacific's efficacy. When Roberts became the railroad's chief engineer in 1870, Cooke wrote to him: "I consider you an *HONEST man* as well as capable!" (emphasis in original). *Courtesy Burlingame Special Collections Library, Montana State University–Bozeman.*

George W. Cass, ca. 1875, president of the Northern Pacific Railroad and a West Point graduate (1832) who entered the Corps of Engineers. He was highly regarded by Andrew Jackson. A tough but fair-minded businessman, in late 1872 he was entrusted with the fate of a virtually bankrupt, poorly constructed system. That the NPRR survived the Panic of 1873 was entirely due to Cass's leadership and skills. *Courtesy Minnesota Historical Society, St. Paul.*

Charles Loring Brace was a prominent philanthropist, founder of the Children's Aid Society, and a leading travel writer. A series of articles he wrote for the *New York Times* in 1873 about the Northern Pacific were likely intended for a book. *Courtesy National Archives and Records Administration, Washington, D.C.*

Capt. Augustus W. Corliss, ca. 1890, was a member of the Eighth Infantry. He left a detailed, handwritten description of the Yellowstone survey, although the last part—with an unknown number of pages—is missing. *Courtesy U.S. Army Military History Institute, Carlisle, Pa.*

William F. Phelps, ca. 1880. A protégé of William H. Seward, Phelps was known for his educational policies, and he became a leader of the normal school movement. The president of Winona College (Minnesota) in 1873, for reasons unknown he was on the 1873 survey as far as the Yellowstone River. Author's collection, from *Northwest Magazine*, 1887.

Lt. Col. Frederick Dent Grant, ca. 1872, while at West Point. After his first cousin, Lt. Lewis Dent Adair, was slain by a raiding party led by Hunkpapa Sioux war chieftain Gall, the president's son was determined to go on the 1873 survey. *Courtesy U.S. Military Academy, Archives and Special Collections Department, West Point, N.Y.*

Rain-in-the-Face (ca. 1835–1905). Like Sitting Bull and Gall, Rain-in-the-Face was also a Hunkpapa Sioux. His name referred to his being splattered in the face with the blood of an enemy during his first fight. Rain-in-the-Face appears to have led the half-dozen warriors who killed Baliran, Honsinger, and Ball on August 4, 1873. *Courtesy National Archives and Records Administration, Washington, D.C.*

Daniel H. Brush, as a West Point cadet. *Courtesy U.S. Military Academy, Archives and Special Collections Department, West Point, N.Y.*

"Heart River," sketch by Charles "Shorty" Graham, 1872. *Courtesy of Special Collections Department, Manuscripts Division, University of Virginia Library (Papers of Thomas L. Rosser).*

Major Robert E. A. Crofton. *Courtesy U.S. Army Military History Institute, Carlisle, Pa.*

Joel Asaph Allen (1838–1921), head of the Scientific Corps that accompanied the 1873 expedition. *Courtesy of Wikimedia Commons.*

Bloody Knife was the lead scout with Custer on August 4. A favorite of Custer's, he was one of the casualties at the Little Bighorn. *Courtesy of Wikimedia Commons.*

Sitting Bull (1834–1890). This photograph vividly captures Sitting Bull's determination and intransigence. A Hunkpapa, he took the lead in rallying Arapaho, Cheyenne, and Sioux warriors against the 1872 and 1873 Yellowstone Surveying Expeditions. The traditionalist Sioux and Cheyenne were attempting to live according to custom; in 1873 Sitting Bull was not looking for a major clash with Stanley's force. *Courtesy National Archives and Records Administration, Washington, D.C.*

Gall, or Pizi (1840–94). A Hunkpapa Teton Sioux, Gall called himself Man-Who-Goes-in-the-Middle, a reference to being in the center of the fighting. Brought up by Sitting Bull's family after his parents died, he was one of Sitting Bull's closest associates in the 1860s and 1870s. A renowned warrior, Gall played a leading role in the 1872–73 fighting. *Courtesy National Archives.*

Northern Pacific Railroad construction. *Courtesy of Montana State Historical Society.*

Hailstone Creek. The embankment on the left side of the railroad bridge was built by the Northern Pacific in the 1940s, but in 1873 the surveyors and the military escort found themselves seeking shelter on the slope of that hill when a hailstorm struck, destroying all their wagons and injuring every soldier and surveyor. By way of perspective, from the top of the trestle to the steam's bed is some seventy-eight feet. (The bridge was completed just after World War II.) *Author's photograph.*

Entrance to the North Dakota Badlands, close to today's Interstate 94 rest stop near Medora, N.D. The drawing, taken from an 1880s issue of *Northwest Magazine*, accurately reflects the splendor of the scene. *Author's collection.*

(*above and opposite*) Eagle's Beak overlooking the Yellowstone near Glendive. One should also notice the sandbars in the river, the same type of flow that Custer encountered August 10. Photograph by William Pywell, 1873. *Courtesy National Archives.*

Looking up the Yellowstone River near the site of the August 11 fighting. The Sioux and Cheyenne had scouts on top of the ridge to the left and were camped behind it. Stanley's infantry spent the night of August 10–11 camped some two miles downriver from the hill where this photo was taken and never saw the smoke from Custer's camp, which was across from the southernmost point of the ridge. *Author's photograph.*

The *Josephine*, named for General Stanley's older daughter, was one of the upper Missouri and Yellowstone Rivers' newest, lightest, and fastest steamers in 1873 and used extensively on that year's Yellowstone survey. *Courtesy of the Overholser Historical Research Center, Fort Benton, Mont.*

Pompey's Pillar photographed by William Pywell in 1873. The members of the Yellowstone survey were camped opposite the famous outcrop. After a half-dozen Sioux fired at soldiers bathing in the river, no attempt was made to cross. *Courtesy National Archives and Records Administration, Washington, D.C.*

Camp from above the Musselshell River in late August. Photograph by William Pywell, 1873. *Courtesy National Archives and Records Administration, Washington, D.C.*

Campsite. Photograph by William Pywell, 1873, likely taken between the Yellowstone and Musselshell Rivers. *Courtesy National Archives and Records Administration, Washington, D.C.*

Panning for gold on the Musselshell River. Photograph by William Pywell, 1873. *Courtesy National Archives and Records Administration, Washington, D.C.*

Wagon train behind a large rock formation between the Yellowstone and Musselshell Rivers. In the middle of the image a long chain of wagons may just be seen. Photograph by William Pywell, 1873. *Courtesy National Archives and Records Administration, Washington, D.C.*

Wagon train near the divide of the Yellowstone and Musselshell. Photograph by William Pywell, 1873. *Courtesy National Archives and Records Administration, Washington, D.C.*

CHAPTER 9

Return to Bismarck and Fort Lincoln

With the fighting behind them and the line of survey completed from Lake Superior to Puget Sound, the 1873 Yellowstone Surveying Expedition began the final stages of its work. It marched from Pompey's Pillar to the Musselshell River (known for its diuretic qualities), where it found medium-sized herds of buffalo. As the surveyors mapped some one hundred miles east to the Musselshell's great bend to the north, two scouts (Charley Reynolds and Gillman Norris) were sent to Fort Benton with letters and telegraphic messages. At the Musselshell's bend, the column split, with Stanley cutting east to the Yellowstone while Custer paralleled it, not turning east until roughly opposite Stanley's Stockade.

The two columns met on September 9. Ferrying operations by the *Josephine* began the next day, with most of the infantry, cavalry, teamsters (and their wagons), and surveyors marching to Bismarck and Fort Abraham Lincoln. However, numerous companies of Eighth and Ninth Infantry slated to return to Omaha stayed aboard the *Josephine*, only to find that the Yellowstone had fallen so fast that the steamer could make no progress (see Bradley's diary).

This final chapter includes a fascinating letter from Konopicky to his patron complaining bitterly about other Scientific Corps members, a *New York Tribune* article (perhaps written in part by Barrows),* a long article including an interview with Custer in the *Bismarck Tribune*, and the *Army and Navy Journal*'s report of the expedition's success.

---

*It is difficult to believe that Barrows never wrote a final letter from the field. More than likely he did write something, but with the 1873 Panic under way, the *Tribune*'s editors saw fit not to publish it.

However, no writing better summarized the problems still facing the Northern Pacific than an editorial in the *New York Times* on September 11 which stated in part:

> If several thousand of our best soldiers, with all the arms of the service, under some of our most dashing officers, can only hold the ground on their narrow line of march for 150 or 200 miles west of the Upper Missouri, what will peaceful bodies of railroad workmen be able to do, or what can emigrants accomplish in such a dangerous region?
>
> The truth is we have now struck upon the last and bravest of the old Indian races of the plain. . . . He is the true specimen of the "noble savage"—brave, treacherous, untiring in war and the hunt, with unconquerable pride, and filled with undying hate of the whites. . . .
>
> But first of all, probably, [the Sioux] must be taught by force. They must learn to submit. In this work the Yellowstone Expedition will not be without its result.

Happily for the returning surveyors, they were unaware of Wall Street's increasing financial turmoil. Then, a week later, with the railroad's surveyors having barely reached Bismarck, word came of the failure of Jay Cooke & Co., the Wall Street crash that set off the Panic of 1873, and the legal bankruptcy of the Northern Pacific. Two editorials (*Sioux City Journal* and *Philadelphia Evening Bulletin*) give differing perspectives of Cooke's failure.

On September 9 the popular "Eck" Eckelson, who had replaced Rosser in the field, wrote to Rosser reporting on the second half of the survey, the letter likely not reaching Rosser until after the Panic had begun. The last entry, taken from the *Army and Navy Journal*, gives what may be considered the definitive account of Indian losses. For better or worse, this article closes with a David Stanley quote of Yogi Berra–like insightfulness.

Konopicky to Dr. Franz Steindachner

September 12

Most Esteemed Herr Doctor:

... We are back on the Yellowstone at the same spot where we first reached it. Next it's back to Fort Rice, quite honestly, to my great joy. I have had enough of this trip and its unpleasant aspects of which I had no idea. Certainly, the most unpleasant aspect was the company I had to live with for months, "The Scientific Corps," and at the top of the list, Professor Allen! Perhaps this astonishes you, but I am also astonished to see the change in the person I had known in Cambridge [Harvard] for two years. The first days out from Fort Rice, he seemed to be very diligent, but as the terrible heat set in, he thought he had sufficiently satisfied things if he [dissected] a few shot birds, in which task he was helped by the general assistant Bennett, who was with us in Cambridge.

The heat tired him so that he had to take a nap every afternoon. Apparently, he thought that others should take up the slack for what he neglected, and to this end he tried to order the photographer Pywell and me around. At first I took it for a while, but then I let him know very clearly that all he had to be concerned about was his own work. I discovered during the trip that the scientific knowledge of this Cambridge-trained gentleman does not extend a jot beyond his birds. He understands nothing about collecting. For example, during the whole trip he was looking for fossilized plants without success. I found many of them, but left them in place to sketch them and this gentleman dares to pretend that he is a botanist and mineralogist, and lets himself be called a professor. Yes, and the best one is that he lets the newspapers call him the manager of the Expedition. But, enough of this gentleman and on to my small sketches of the trip. On July 26th we crossed to the far side of the Yellowstone after crossing a desolate, barren, hilly prairieland and jagged sand mountains—these are the main features of the country we traveled through. It went on without change in terrible heat until August 4. That day we had a skirmish with Indians. We were about to descend from the high prairie into the Yellowstone Valley to make camp after a tiring march, when we saw a large part of the Valley in flames. Apparently, the Indians who knew our plans, wanted to burn the food for our animals. We immediately climbed back up and found the first two victims from our company in the grass, shot through the chest. They were [a] cavalryman and a sutler. But see earlier letter [about

Baliran] and the Doctor who had ridden a little too far ahead. Another cavalryman was also missing; we found his skeleton on our way back [last week]. The same day, some 70 cavalry, who rode ahead of the column, had a longer battle with some 200 Indians.

The next encounter was on August 11 with several hundred Indians pursued by 500 cavalry under General Custer. After a long battle, our losses were one man killed, one officer seriously wounded, six men with minor wounds and 3 horses dead. This day, we also used our two canons, which I witnessed. General Stanley ordered the canons to shell the Indians camped on the other side of the river. The Indians also attacked once at night. They stampeded our cows and we lost 40 of them.

From here we went northwest to the Musselshell River, a small but very picturesque river, which we reached in three days. This was an interesting part of the journey because we saw many herds of buffaloes, one of them of 2,000 to 3,000 head. Many of these magnificent animals were killed, and I succeeded in killing one too. You may be amazed. Herr Doktor, I hope not too much, although I have to admit that the animal was already wounded and sitting on the ground. It took five revolver shots to put him down. We ate a lot of buffalo meat, but also mountain sheep, antelopes, elk, etc. Of all of these sorts of animals, three grizzly bears were also killed! Allen and Bennett shot none, despite the fact that they rode around with two flintlocks, revolvers and knives like walking hardware stores.

I . . . now have 60 sketches. . . .

Your Deeply Indebted, E. Konopicky

Bradley's Diary, August 19–October 1

AUGUST 19TH. Marched about 6. In advance. Country rolling and descending all day. Killed several buffalo on the march. Found plenty of water in holes, result of recent rains. Reached the Musselshell and camped at 3 P.M. We followed Baker's [1872] trail mainly from the Yellowstone. Found the road generally good; made 40 miles westing [*sic*], and 20 marching. Whole distance across on this trail, 60 miles. The distance between the two streams at the nearest point cannot be more than 20 miles.[1] Baker's Trail is probably the best point to cross. To the

---

[1] The minimum distance between the two rivers is some fifty miles, though the rivers roughly parallel each other.

east of this the country is very rough and judged impassable. The location of the Musselshell on the maps is very faulty.

AUGUST 20TH. Marched at 6 A.M. Rear guard. Forded the Musselshell and followed the north bank, course nearly due east. Country rolling and broken, road very good. Rained all the morning and very cold. Heavy growth of pine in sight all day. Camped on the river at 12. Distanced marched 14 miles.

AUGUST 21ST. Marched at 9 A.M. In advance. Waiting for the Engineers to complete yesterday's work. Marched down the valley and found the road bad. Camped at 1 P.M. Distance 5 miles.

AUGUST 22ND. Marched at 6 A.M. Rear guard. Forded the Musselshell six times, then left the valley and found trail over the bluffs. Had very broken country all day; timber plenty on the river and hills. Passed an immense coal field in the afternoon. [Mule] teams breaking down fast from hard work and short feed. Camped on the river at 4. Distance 20 miles.

AUGUST 23RD. Marched at 8 A.M. In advance. Kept in the valley and crossed the river four times before noon. Very hot and dusty. Camped at 1 P.M. in a beautiful place on the river. Plenty of trees and fine grass, the best camp we have had on the march. Heavy timber, ash and cottonwood on the river and fine pine on the hills. Distance marched 11 miles.

AUGUST 24TH. Lay in camp all day to rest the stock. Game plenty on all sides of us.

AUGUST 25TH. Marched at 6 A.M. Rear guard. Had good road down the valley and marched fast. Forded the river twice. Middle of the day very hot. The river bears to the north as though making big bend.[2] Camped at noon. Distance 14 miles. Distance from first camp on Musselshell to big bend 64 miles.

AUGUST 26TH. Moved camp over to the east bank of river. Engineers out with Cavalry escort to examine the country east to determine our future line of march. If the country to the east is impracticable, we return to the Yellowstone. The valley of the Musselshell is very pretty though narrow, the average width not over a mile. The soil is good and grass fine; cottonwood timber is very heavy on the stream and the hills are full of pine.

---

[2] North and east of today's Melstone, Mont.

AUGUST 27TH. Marched at 6 A.M. Rear guard. Left the river and marched a little north of east over a rough country, dry and barren, no trees or grass. Marine fossils in abundance. Camped at 12 on small creek running to the west: little grass or water. Distance marched 13 miles.

AUGUST 28TH. Marched at 5:25 A.M. In advance. Clear and cold, first cold morning of the season. Route lay over a miserably poor country, rough and utterly barren. Custer in advance to look out road, led us too far north and had to turn south again. Country full of marine fossils. Camped at noon on unknown creek, supposed to be a branch of the Big Porcupine. Wind blowing a gale and dust in everything. Distance 16 miles. Engineers and escort did not get into camp being on a line south of us. Burned signal fires and threw a number of shells to guide them, but they did not see them. Sent out a scouting party too, but they saw nothing of them.

AUGUST 29TH. Marched at 5:30 A.M. Rear guard. Cool morning, route south to east. Found the Engineers and escort about 9. They bivouacked about 8 miles from our camp and supped on bear meat. Custer started with 6 companies of Cavalry to take the Engineers across the country to the mouth of Glendive, it being considered unwise to take the whole command across. The main column marched down the west side of the Big Porcupine: found good road and camped at 3 P.M. Distance 18 miles.

AUGUST 30TH. Marched at 8 A.M. In advance. Left the Porcupine and struck off southeast for the Yellowstone. Stanley was a poor pathfinder today. We marched all day without water and bivouacked at 8 P.M. on a bare hillside. No water or wood to be had and no one knows how far it is to the Yellowstone. Mules are giving out fast; 10 shot today. Unable to travel any farther. Some teams did not get into camp till midnight. Distance 26 miles. Passed over a fine fossil region nearly all day: great numbers of Ammonites,[3] [unclear], and huge Sea Turtles. Some of the Turtles near perfect in form and 12 and 15 feet long.

AUGUST 31ST. Marched at daylight. Rear guard. Followed the divide down toward the Yellowstone. Very hot day and much suffering for want of water. Sent out parties to search for water, but found none till we struck the Little Porcupine about 11, 30 hours from the last water. A good many animals broke down. Camped on the Yellowstone at 1:30.

---

[3] Ammonites are circular seashells. The word is derived from the Latin word for "ram's horn."

Distance 14 miles. The country between the Musselshell and the Little Porcupine is the poorest I ever saw, good for nothing but fossils.

SEPTEMBER 1ST. Marched at 5:30. In advance. Had a good road down the valley and the country on the river looked very nice especially about the mouth of the Rosebud [on the other side of the Yellowstone]. Camped at 1 P.M. Distance 13 miles.

SEPTEMBER 2ND. Marched at 5 A.M. Rear guard. Had to climb the "big hill" the first thing.[4] Took the train 5 hours to get up with help of the troops. Made the same march as on the 5th of August and camped in the valley at 3. Distance 10 miles. All the bridges built on the march up the valley carried away by heavy rains since we passed a month ago.

SEPTEMBER 3RD. Marched at 5:30 in advance. Rain commenced falling just as we broke camp and continued all the forenoon. Cold and disagreeable enough, very muddy and the Alkali soil stuck to the feet till they were twice the usual size, which made pretty hard marching. Camped at 11 A.M. at the place where the men of the 7th Cavalry were killed on the 4th of Aug. Distance 8 miles.

SEPTEMBER 4TH. Marched at 5:45. Rear guard. Went over the route of the 4th of Aug. Fine day and pleasant marching. Found 2 of Yates' horses lost last month. Camped at 2 P.M. Distance 18 miles.

SEPTEMBER 5TH. Marched at 5:20 in advance. Followed the valley for 9 miles then struck out over the bluff to avoid obstacles. Camped on the river at the point where the *Josephine* left us in Aug. Showers nearly all day. Distance 14 miles.

SEPTEMBER 6TH. Marched at 5:45. Rear guard. Fine cool morning. Marched down the valley to the mouth of Custer's Creek and turned up that stream. Good water in holes all along the creek. Camped at 12:30 in the Bad Lands. Distance 16 miles.

SEPTEMBER 7TH. Marched at 5 in advance. Fine cool day and good marching. Took the old trail through the Bad Lands and reached the old camp on Bad Route Creek at 3 P.M. No water in the creek. Found some water holes a few miles east. Distance 21 miles.

---

[4]The U.S. Geological Survey today refers to this boomerang-shaped butte as "Big Hill." It is opposite Miles City.

SEPTEMBER 8TH. Reveille at 3 A.M., but did not start until 7 on account of the mules being scattered in search of water. Some of them were not found at all. Had a good march and camped on unknown creek at 2. Found good water and grass. Distance 14 miles. Found Capt. French [7th Cavalry] waiting for us with grain and mails.

SEPTEMBER 9TH. Marched at 5:10. In advance. Followed the old trail for a few miles then took a cutoff the Cavalry had made, shortening the distance to the Stockade. Warm day and good marching. Camped opposite the Stockade at 1 P.M. Found the steamer *Josephine* here, just [having] arrived. Distance 19 miles. From the big bend of the Musselshell to the crossing of the Yellowstone 220 miles.

SEPTEMBER 10TH. Received orders from Gen. Terry relieving my Battalion from further duty with the Yellowstone Expedition and directing me to proceed to Omaha on steamer *Josephine*. Steamer commenced crossing the Cavalry over the river which is too deep to ford and but for the steamer, we should have been puzzled to get across.

SEPTEMBER 11TH. Cavalry all across and train crossing today. Everything working well and everyone in a hurry to be off for home; one-half the train will be over tonight. Had one man badly hurt loading wagons.

SEPTEMBER 12TH. Cold morning. Custer started off for the Little Missouri with 6 companies Cavalry as escort to the Engineers. All the train over tonight except the Headquarters teams.

SEPTEMBER 13TH. Broke up camp on the north bank of the Yellowstone at 6 A.M. and just the Battalion on the steamer for home. Crossed the river to the Stockade and took on board 100 tons Quartermaster and Commissary stores with all the wounded. sick and discharged men of the Expedition to convey them to Fort Lincoln. Lay at the Stockade all night ready for an early start in the morning.

SEPTEMBER 14TH. Started down the Yellowstone at daylight. Got aground within a mile of the Stockade and lay about 3 hours before the boat floated. Struck a bar again about 9 and were hard and fast—could not start her. Landed the men to lighten up and sent rations and blankets ashore for the companies. Capt. Porter, Lieuts. Griffith and Jones sent ashore to remain with the Battalion. Worked hard at the Steamer all day and evening but only moved her a few feet.

SEPTEMBER 15TH. Hard aground this morning. Succeeded in getting off the bar about 4 P.M. Took the men on board and started. Run till night and tied up just below the mouth of Glendive Creek. Capt. [Grant] Marsh of the steamer reports the river falling at the rate of 1½ inches an hour and doubts whether he can take her out with her present load.

SEPTEMBER 16TH. Landed the men at daylight to lighten the boat over Glendive Falls,[5] Captain reports but 2 feet of water on the rapids and the boat drains 32 inches. Landed 12 tons of Quartermaster and Commissary Stores consisting of Horse and Mule shoes, wagon grease, vinegar, pickles and sauerkraut and abandoned them to enable the steamer to clear the rapids. This with the men off makes 50 tons difference in load. Got over the Falls about noon and made 25 miles before night. Tied up at dark.

SEPTEMBER 17TH. Aground again this morning. Landed the Battalion to march around Coulson's Island. After some hours work without starting the boat Capt. Marsh reported that he could not get out of the Yellowstone unless the Steamer was lightened up 5 or 6 inches. A Board of Officers consisting of Col. Townsend, Major Lazelle, Capt. Owens and Lieut. Pease were ordered to report on the matter. [They] advised that all the Commissary Stores on board the Steamer be abandoned some 60 tons in all to enable her to proceed. Ordered the stores landed and worked all night at them. Had them piled in the woods about 1 mile above the head of Coulson's Island and covered with [tarpaulins].

SEPTEMBER 18TH. Finished landing stores this morning and started down steam. Stuck again in less than 15 minutes and were fast. The Battalion marched down the west bank and waited for the boat at the head of the island. Worked hard all day, but could not get off the bar. Ordered the Battalion to return and bivouacked on the bank opposite the Steamer. Sent rations and blankets ashore.

SEPTEMBER 19TH. Anniversary of Chickamauga.[6] Got the Steamer to the east bank this morning and landed the rations and ammunition to try again to get over the bar. Succeeded after 5 hours hard work, reloaded

---

[5]Actually rapids.
[6]Leading his infantry brigade on horseback and advancing across an open field, Bradley was hit in the chest. He recovered in time to participate in the Atlanta and Nashville campaigns.

ammunition, etc. and dropped down to Coulson's Island after grounding a couple of times more. Tied up at 6. Rained all the early part of the night.

SEPTEMBER 20TH. Had a good run today, no trouble till near night. Stuck on a bar in Ashbend at sundown and did not get up to the bank till about 9 o'clock. Run about 40 miles today, the best since we started.

SEPTEMBER 21ST. Started soon after daylight and after running 2 or 3 miles, came on the shoalest [sic] place in the river, 18 inches of water. Unloaded stores, ammunition, baggage, etc. and landed the men, lightening up to 18 draft, but it took until the middle of the afternoon to get off. Run till dark and tied up for the night.

SEPTEMBER 22ND. Got under way at daylight. Passed old Fort Gilbert at 7 A.M. 26 miles from Fort Buford. Struck a bar at 10 and had to land the troops. Got off in about an hour and proceeded. Reached the mouth of the Yellowstone and passed into the Missouri at 1:30 P.M. 9 days from the Stockade—130 miles. The Yellowstone below the Stockade is well wooded and has a good deal of fine country on it. The channel is changeable and uncertain and the stream widens in places so as to make navigation almost impossible at time of low water during the summer months. . . . Reached Fort Buford at 2 P.M. and went up to the Post to see [Col.] Hazen.[7] Spent an hour with Mrs. Hazen and had a pleasant time. Buford is a good looking, 6 company Post with new and comfortable quarters. The country around it is pleasant and it commands a fine view of the Missouri and Yellowstone. Left Buford at 4:30 after drawing rations to take us to [Fort] Lincoln. Went down stream fast and tied up for the night about 9.

SEPTEMBER 23RD. Had a pretty good run today. Aground but once. Rained all the afternoon and the boat was pretty wet. Tied up at 9.

SEPTEMBER 24TH. Rained all day. Aground several times and lost a good deal of time. Tied up soon after dark.

SEPTEMBER 25TH. Raining again today. Slow running on account of low water. Reached Fort Stevenson at 2 P.M. The Post is a pleasant looking one and well built. . . .

SEPTEMBER 26TH. Raining most of the day. [Low] water shoal and the boat aground several times. Made about 50 miles and tied up at 6 P.M.

---

[7] The two had served together in the Civil War.

SEPTEMBER 27TH. Had a good run and reached Fort Lincoln at 11:30 A.M. Found Stanley and all the Expedition there before us having marched 250 miles across the country while we were making 400 by steamer. . . . Went over to Bismarck in afternoon with Lazelle, Burt and Pollock to go to St. Paul.

SEPTEMBER 28TH. Lay over in Bismarck, a dirty little border town.

SEPTEMBER 29TH. Started for St Paul at 7 A.M. and reached Fargo at 7 P.M. Took supper at Morehead and left there at 9, on stage for Breckenridge [Minnesota]. Rode all night.[8]

SEPTEMBER 30TH. Reached Breckenridge on the Red River at 4 A.M. Started at 7 [A.M.] for St. Paul, had a pleasant ride and reached St. Paul at 8 P.M. Minnesota ought to be known as the Lake State from the Red River to the Mississippi you are scarcely ever out of sight of a lake. There are said to be 20,000 lakes in the state.

OCTOBER 1ST. Very beautiful day. Called on Gen. Terry in the morning and dined with him in afternoon. Had a pleasant visit. Am very much pleased with St Paul.

OCTOBER 2ND. Started for Sioux City at 8 A.M. [Rail]road runs down the valley of the Minnesota through a beautiful country and well settled. Fort Snelling, at the junction of the Mississippi and the Minnesota [Rivers], is a pretty place. Reached Sioux City about 10 P.M.

New York Tribune

THE YELLOWSTONE EXPEDITION

ARRIVAL OF GENERAL CUSTER AT THE YELLOWSTONE STOCKAGE— SUCCESS OF THE EXPEDITION

NEW YORK, SEPTEMBER 21.—Advices from the Yellowstone Expedition under date of September 6, by way of Bismarck, D.T., Sept. 16, state that

---

[8]Nothing better illustrates the logic-defying Northern Pacific decisions; in this case failing to complete the Brainerd to Minneapolis–St. Paul spur. Rather than spend an extra day going from Fargo to Duluth to St. Paul, travelers like Bradley transferred at Fargo for a Breckinridge stagecoach, from there taking a train to Minneapolis–St. Paul.

Gen. G. A. Custer, with three squadrons of the 7th Cavalry, comprising A, B, C, D, E, G, and M Troops, together with the engineers, reached the stockade on the Yellowstone River Sept. 4, having separated from the main command three days' march east of the Musselshell River. General Stanley, with the Infantry, one squad of cavalry, and the train struck directly for the old trail on the Yellowstone, and are not expected at this point until the 12th.[9]

The expedition has become a complete success, every object for which it was inaugurated having been effectually accomplished. An area 20,000 square miles of unknown country to the south and west of the Yellowstone has been explored, the Indians thoroughly whipped, and the survey of the Northern Pacific Railway completed. The route selected by Mr. Eckelson, chief engineer of the surveying party, will follow the east bank of the Yellowstone as far as Tongue River,[10] crossing at that point, thence a short distance along the west bank to the Big Porcupine, crossing the divide in that vicinity of the big bend of the Musselshell, following the valley of this river to a connection with Baker's survey of last year.

Gen. Custer made a direct trail to the stockade, after separating from Gen. Stanley, across a terrible country, almost destitute of water and vegetation, accomplishing the march of 140 miles in five and a half days, with a train of 30 wagons, and with the loss of just six animals. No trouble was had with the Indians since the battle of the Big Horn; the wounded are doing admirably, and are considered all out of danger. . . .

Published September 22

FORT LINCOLN, D.T., VIA BISMARCK, D.T., SEPTEMBER 29.—The steamer *Josephine*, with 8th and 9th Infantry in command of Gen. Bradley, arrived today from the Yellowstone. This is the last of the three divisions of the expedition which separated on the return from the Yellowstone. The *Josephine* had expected to make the trip in four days, but having on board a large amount of the surplus stores taken from the stockade, and the water in the Yellowstone falling rapidly after her ascent, she struck on several sandbanks and did not reach the mouth of

---

[9]It is not known whether Barrows wrote any or all of this, or if Barrows's last correspondence was edited down to this short passage.

[10]In the early 1880s the Northern Pacific was actually built on the Yellowstone's eastern side.

the Yellowstone until the ninth day after leaving the stockade, and Fort Lincoln on the fourteenth day. They were obliged to leave 70 tons of freight at a point 60 miles from Buford. General Hazen will send from Buford for most of these stores. A company of the 6th Infantry was left at Fort Buford. The 8th and 9th go to Omaha on the *Josephine.* The 22nd Infantry, with them General Stanley, leave for Forts Sully and Randall tomorrow by boat. The Cavalry remains at Fort Rice and Lincoln. The Scientists have left for the East.

Published September 30

Bismarck Tribune

HOME AGAIN

RETURN OF THE GREAT YELLOWSTONE [EXPEDITION] STANLEY AND CUSTER ENCAMPED AT FORT LINCOLN THE TROOPER GENERAL INTERVIEWED— WHAT HE SAID

BISMARCK, D.T., SEPTEMBER 24—"Custer has arrived" was passed from lip to lip on the street on Monday evening [September 22] and the subject at once became the theme of general conversation. If the General is as successful at taking his enemies by surprise as he was in surprising his friends on this occasion, he must often catch them asleep. We knew the General in the army,[11] and knew him to be a man in whom sloth took no delight, and as an officer, the hero of more rapid marches and harder fights than almost any other noted in history, and for the last few months he had almost literally kept a moving circle of troopers around the working force and Infantry of the great Yellowstone Expedition; but for all this we were not prepared to witness quite so much energy as was shown in this, the dying kick in the whole transaction.

We were waiting and waiting for the steamer *Josephine* and its load of "dough boys." That the cavalry would out-march a steamboat, and that

---

[11] Clement Lounsberry (1843–1926), the *Bismarck Tribune*'s publisher, was colonel of the Twentieth Michigan Infantry during the war. This helps clarify that it was he, not his assistant, Mark Kellogg, who wrote this.

downstream on a rapid river, was not even thought of. So when the news was first circulated upon the street, the question asked was,

"Why, how did he come?"

"Marched of course."

"But it isn't possible."

"Nonsense, nothing is impossible with Custer."

And so it proved; the cavalry had certainly and with it the necessary train of supplies.

THE TRIBUNE ON HAND

Bright and early the following morning our reporter was astir. The *Tribune* man never sleeps; only one of him at a time at all events. Kindly placed under obligations by Commodore [William J.] Kountz, the reporter accompanied an excursion party under the guidance of that gentleman, stepped on board an extra [sic] and soon arrived at the landing, where the steamer *May Lowry* transported all hands to the opposite shore, from whence an occasion upon the General and his encampment were planned. The encampment was easily found, just below Camp Gronewald,[12] the proposed winter quarters of the expedition. The systematically arranged "dog tents,"[13] the long lines of horses, the lolling, listless, blustering, swearing, good natured and jolly crowds of men, carried us back in imagination to the days when we, too, "slept on the tented field" (and P.S., don't want to do it anymore).

Quickly finding his way to the General's tent, our reporter stepped in doffed his beaver [top hat], done his level best in the regulation military salute, presented his credentials and received a hearty greeting. We are no Jenkins;[14] but nothing in the panorama outside so vividly recalled old times as did the dress of the man before us.

It was the very same uniform, in style, the appearance of which, clothing the same dashing cavalryman, had so often startled the corn-fed Jonnies by the yells of welcome it called out from the boys in blue. Custer had arrived the night previous, of course, in the frontier regulation buckskin, but this had been doffed, and in its place donned the well remembered plain blue pants, with their yellow stripe, the blue flannel shirt, with its wide collar and cuffs, loose at the throat, the black slouch

---

[12]Unclear as to location unless it was an inside joke.

[13]"Pup tents."

[14]Reference unknown.

hat, and the jingling spurs. The same long curling hair and the accompanying mustache make up in brief the General's *personnel*, and it will all be sufficient for those who have once seen it. We were directed to a seat, and after further greeting and a little general conversation opened up our batteries as follows.

Reporter.—You are of course aware, General, that one of the first duties of a man who has finished up some great undertaking, has met with some sudden turn of the wheel of fortune or misfortune, or from any cause has become suddenly conspicuous, is to undergo a thorough pumping at the hand of a myriad of impecunious newspaper reporters.

Gen. Custer.—Oh, of course; but so long as it is a duty, of course, it is all right. I am pretty busy today, so soon after reaching camp, but any information I may be able to give you will be given with pleasure.

Reporter.—If you will [be] kind enough then, General, you may give to me, any items of interest in connection with your trip which you deem of interest to the public.

Gen. Custer then proceeded (we shall not attempt to follow his language) to suggest that the advance of the expedition to the Musselshell was probably pretty thoroughly written up already. The only item of interest, indeed, to the general reader, was the battle with Indians under Sitting Bull. This was of course also understood, as Gen. Stanley's report covered the details. At the time the attack was made, the command was at a halt, the General and his staff resting in the shade of a tree, the former with his boots off. The first general notice of the presence of the Indians consisted of a volley from the foe, which had rather a tendency to wake things up. The men and horses were scattered around somewhat promiscuously, but a very short period of time sufficed to rally them and the line was at once formed. After a short skirmish the enemy were driven across the Yellowstone, and the rest is known.

After the battle Stanley's march was resumed to the head of the Great Porcupine, which was to be the turning point of the expedition. The Indians were evidently well punished as not a single top-knot was seen afterward.

After reaching the head of the Great Porcupine, the return march was at once commenced, and the cavalry soon lost sight of Gen. Stanley and the infantry. The march through from this point to the mouth of Glendive Creek where the crossing of the Yellowstone was to be made was without interest of note or difficulty. Good camping grounds, with plenty of grazing and good water were found along the whole distance. Game was plentiful, antelope, black-tail deer, and the feathered tribe being in abundance. Reached the Yellowstone on September 9th marching at the rate of 30 miles per day.[15] Upon arrival at the Glendive, the command encamped and an escort with supplies were sent back to meet the infantry. Five days thereafter the Infantry arrived and on the same day the steamer *Josephine*. The *Josephine* at once proceeded to [ferry] the expedition [to the east shore], and to take on board the 8th and 9th Infantry, which she will carry through to their destination and winter quarters at Omaha.

Here Custer again left Stanley rearwards, and resumed his march for Fort Lincoln. Leaving the old surveyed route, the one passed over on the outward march, the command took a more direct course, avoiding part of the Bad Lands, and finding fine streams of clear running water, better grazing, and a good road clear through. Gen. Stanley with his infantry and the main wagon train are expected to arrive here in a day or two. It is generally understood that the Infantry command not otherwise assigned will winter at Fort Rice.

The General states that himself and his command enjoyed exceedingly good health during the time they were out, The trip in from Glendive was made in the remarkable short time of eight days, the last day's march being forty miles.

Lt. Braden, who was so seriously wounded in the fight with Sitting Bull, and who was reported past hope, is now on the *Josephine*, and the General says recovering fast. . . .

The General brings in with him a young menagerie, which he proposes to send to Central Park, New York. This "magnificent aggregation," as Barnum would say, consists in part of wild cats, porcupines, antelope horns, petrified wood, and more wild cats (also bob-tailed).

The men and animals all show service, and are doubtless prepared to appreciate the rest in store for them; and who will say they have not earned it.

ONE DAY LATER

General Stanley with his infantry and trains has arrived, and they have certainly earned a reputation for rapid movements.

The only portion of the expedition now not in is with the *Josephine* and why the steamer has not reached its destination is the great conundrum.

---

[15]Actually September 4. There is a factual and compositional sloppiness reflecting the understaffed *Tribune's* rush to set this interview in type.

That it has met with some accident or detention is certain, but what that might be is certainly only a matter of speculation.

Rosser's Success

THE YELLOWSTONE EXPEDITION

General Rosser, in charge of the railroad survey of the Stanley Yellowstone expedition, has submitted to the authorities of the Northern Pacific Railroad Company his official reports of results thus far accomplished by the expedition. He finds the new and final route across Western Dakota from the Missouri to the Yellowstone River entirely practicable and satisfactory, it being greatly superior to those of former days. Good water was found the entire distance. Coal outcrops at various points in veins several feet in thickness, and timber is now more abundant than on former routes.

The report states that the main body of General Stanley's expedition accompanied the Scientific Corps, and most of the press correspondents did not accompany the Engineers, who were escorted by General Custer's detachment, but followed the old abandoned route south of Heart River, hence descriptions of the region traversed by the main command do not apply to the country traversed by the new route for the railroad.

The directors of the company have accepted the new line recommended by General Rosser from Bismarck, the present end of the track, to the Yellowstone Crossing, and have called for proposals to grade and bridge this section of 205 miles. The expedition is now prosecuting the survey westward up the left bank of the Yellowstone to Pompey's Pillar, where it will join the survey last year from the west, and thus complete the survey line across the continent.

Sources: Numerous newspapers including The Army and Navy Journal, *August 23, 1873. Likely written by the Northern Pacific's Board of Directors.*

Sioux City Daily Journal

FAILURE OF JAY COOKE & CO.

SEPTEMBER 20.—The failure of Jay Cooke & Co. means the failure of the Northern Pacific Railroad Company. Not only will the line not be extended beyond the Missouri River, as has been announced it would be without delay, but we doubt very much that the line will be continued to be operated to Bismarck.[16] Jay Cooke & Co. have made a desperate struggle to carry through this impracticable enterprise. We do not doubt that, at the start, they were greatly deceived as to the character of the country through which it was proposed to build the road and the prospect of the enterprise proving profitable.

Of late their dependence has been upon deception. Rose colored pictures have been presented of the barren country through which the road has its track, in the hope of making a market for bonds, and of inducing home-seekers to come in and invest in the lands donated by the Government.[17]

The press has been suborned, and Government officers have reported to order, and the expensive expedition to the Yellowstone was figured through in sheer desperation. But all the skill and all the astonishing power of the financial agents have not averted the bursting of the bubble. We have rebelled, and spoken fearlessly and distinctly against the impositions so desperately attempted. We were prepared for the collapse. The only questions were to the time of its occurrence and as to who and how many were to be the victims.[18]

Philadelphia Evening Bulletin

YESTERDAY

SEPTEMBER 19.—The failures of some men are more honorable and more beneficial to themselves than all the successes and prosperities of others. And the failure yesterday of Jay Cooke & Co. was received by the community where the head of that [banking] house is best known, and therefore most honored and loved, with a sense of profound sympathy, so hearty and sincere that the misfortune cannot but be lighted of its greatest burden to those who have it to bear.

Here and there are those who secretly rejoice that their envy and jealousy have at last been ratified; but, like the traitors on the true cause

---

[16]Fargo to Bismarck operations were curtailed, but resumed in 1874.

[17]The first wave of "home-seekers" mostly took title to free government land. When the Northern Pacific could sell land, it averaged some six dollars per acre, usually on installment.

[18]One of the victims was chief engineer Milnor Roberts, sixty-two, whose life savings were almost entirely wiped out.

on the say that Abraham Lincoln died, they timidly hide their feelings under the mask of decent regret, because they know that public sentiment would not brook any open expression of them.

We do not believe that the time has come, yet, to write the history of the house of Jay Cooke & Co. We give elsewhere today, an interesting sketch of his grand career during the war, a career marked not only by a sublimely patriotic confidence, gigantic energy, and statesman-like sagacity, but by the popular test of all merit, a perfect success. And the men who carried without even staggering, the tremendous burden of the national debt, and confronted, without fear, the darkest threatening of the days of the Rebellion, will now find that the country remembers their sacrifices with gratitude and confidence. The time has not yet come for Jay Cooke & Co. to fall. As thousands of hearts beat with sympathy towards them, very rarely drawn out by any simply commercial or financial house, so thousands of strong arms will be stretched out to give that sympathy practical shape and efficacy.

Eckelson's Report to Rosser

Yellowstone River [opposite Stanley's Stockade]
September 9, 1873

Genl. T. L. Rosser
Div. Engr. NPRR
Minneapolis, Minn.

Dear Gen.:
Yesterday the steamer *Josephine* arrived at the stockade and brought all your letters and dispatch. I am very glad the company are so well pleased with the result of your line via Sweet Briar and that they intend building at once. . . .

We commenced work as directed immediately opposite Powder River. Found the valley of the Yellowstone good for about eight. Then ran into cut bluffs.[19] Was [as] bad as those above Glendive, but supposed to avoid the bluffs extending out abruptly into the river. Most of the time we were able to run the line on the edge of the river and in that way were enabled to make good time. And in this way the valley continued all

---

[19] Eckelson means a bluff that has to be cut into, which is normally difficult and expensive.

the way to Pompey's Pillar. The distance as I wrote you was 201 miles from Glendive's and out of that distance we had about 35 miles of those cut bluffs.

We took the south side of the valley just as close [as possible to] the North and find it in every way far superior to the north. *Almost all the way the valley is splendid* until you reach the Rosebud River. There we have an excellent crossing and a good valley on this side (*the North*). Then when we reach the Big Porcupine [Creek] we have excellent alignment and easy grade via the Musselshell River which will be as cheap to construct and cheaper I think than the line up Sweet Briar Creek. I am confident you will be pleased with it.

The Musselshell River [is] beautiful indeed. The valley is all you could ask, and fine pine timber all over the hills [on each side of the] river. The crossings are excellent all of them good hard boulder and gravel bottoms. In the 66 miles that we ran the line [of] the Musselshell we will have to cross about 10 or 12 times. The work will be around 3,000 or 4,000 [dollars] to the mile. And the best of material, the soil is productive and is very available for agriculture.

Leaving the Musselshell we have a good line until we strike the head of the Porcupine River [*sic*] which is about 35 miles out from which point we have one of the most beautiful views that is on the route. The whole country slopes down to the Yellowstone and the creek's course is good running about south and east.

We made every effort to get a good line throughout the proposed route but found it impossible. There is no plateau on the riverside between the Yellowstone and Missouri river, but many a back bone high and ragged—and runs in a zig zag direction north and south making an immense distance. We found trouble to cross this ridge at all.

We traveled 40 miles on the Missouri slope—the country very broken for railroad purposes but a good wagon road. The water was scarce and we just managed to get through. We were compelled to make from 20 to 25 miles per day on the last 130 miles. The main command could not have followed through. We ran the line on through and connected at Glendive Creek and came out perfect. Our map is finished up to the Musselshell. . . .

Gen. Stanley with the command arrived here last night all looking *pretty well played out*. I called on him and asked for escort to go with me at once for the Little Missouri. We cross today and start tomorrow morning. Gen. Custer goes with me and takes six troops of cavalry and 4 of infantry. Stanley will not get away for 7 or 8 days.

I will commence at the Little Missouri and do all I can. I will continue my line all the way down to where we struck Heart River on the Sweet Briar Line, also eastward as far as necessary until I hear from you again. I start with 30 days rations and forage. Stove wood [supply is] fair. I think they will go through it. The boys are all well and ready for [unclear] hard work.

Source: Northern Pacific Railroad, Unregistered Secretary's Letters, Minnesota Historical Society, St. Paul.

Army and Navy Journal

THE BIG HORN FIGHT

NOVEMBER 8.—Colonel D. S. Stanley . . . under date of Fort Sully, Oct. 11, reports the following as the Indian account of the fight on the Yellowstone in August last.

After the return of the expedition, the Little White Swan,[20] a friendly Minneconjou chief, who has lived at the Cheyenne Agency for the last six years, called young Antoine Clement[21] to a feast. The latter was in the fight at [the] Big Horn, a scout on our side. The White Swan told Clement that four young men who participated in the attack had been to see him and said that the Indians lost 4 killed and 12 wounded in the *two* [emphasis added] engagements, besides 26 ponies either killed or dead since; that one of the wounded had since died, and that others would probably die.

The leaders in the attack were "Red Ears" son, a Brulé, and the "Bull without Hair," a Minneconjou. Red Ears draws rations at the Cheyenne Agency. The Indians engaged were 800 to 900 strong with a very few Hunkpapas. "Long Dog,"[22] a Hunkpapa, went to see "Sitting Bull," who

---

[20] Little White Swan (ca. 1838–1900) was the son of the Minneconjou chief of the same name (ca. 1810–77).

[21] Antoine was Basil Clement's son and a half-brother, first cousin, or nephew of one of the White Swans. Antoine's grandfather was a well-known early 1800s trapper and mountain man.

[22] Little is known about Long Dog other than that he fought at the Little Big Horn, survived (another Sioux with the same name was killed), and retreated with Sitting Bull into Canada in 1877. He may have been a brother of Rain-in-the-Face.

refused to join the war party, saying his promise to Pere de Smet[23] was "medicine" and he would not fight unless attacked.

After Col. Custer's movement had drawn them [the Indians] across the Yellowstone to the south side, they found no buffalo, and as the antelope have nearly all died this summer over an unusual extent of country, the hostile camp was compelled by starvation to go to White Clay, Spotted Tail's Agency,[24] for rations, where they are now. The Indians said that they had sent runners to the Oglala, but found the latter away [chasing] the Pawnees. Col. Stanley, in conclusion says: I give this Indian story for what it is worth. My experience is that the Sioux generally give a pretty correct account of their losses. They may, however, conceal them.

---

[23]Father Pierre-Jean De Smet (1800–1873) came to the United States in 1821, was ordained a Jesuit, and began missionary work among the Plains Indians in the 1830s. He played a significant role in the 1868–69 Fort Laramie peace treaty negotiations.

[24]On the Missouri in southwestern [South] Dakota.

Conclusion

The 1873 Yellowstone Surveying Expedition was a definite success for both the army and the Northern Pacific. As the military had hoped, Sitting Bull and his followers were defeated not just once but three times at Mandan and along the Yellowstone. When Custer entered the Black Hills the following summer, no Indians rose to meet him, the balance of power being so heavily tilted in the army's favor.

As for the Northern Pacific, for a few weeks at least, the railroad could bask in the sunlight of its completed line of survey from Duluth on Lake Superior to Tacoma on Puget Sound. With the route known, the railroad's executives assumed, planning could go forward. In fact, they quickly called for and received bids for the 205 miles of construction between Mandan-Bismarck and the mouth of Glendive Creek.

The 1873 survey also proved that a railroad on the western-northern side of the Yellowstone was not feasible. The reason was simple: large portions of the western side were made up of rugged ground and high, difficult-to-pass buttes—Big Hill, opposite Miles City being just one example. Even today, from immediately west of Glendive (Forest Park) to Billings (some 235 miles), not a single community and no fully connected roads (paved or dirt) road exist. This contrasted sharply with the Yellowstone's eastern side, which had, with the exception of easy-to-cross rivers and a handful of outcroppings, mostly rolling hills and bottomlands with easy gradients.

Sadly, Jay Cooke and Milnor Roberts's desire to avoid building on the Crow reservation was stymied by geographic reality. It was not that they were unaware of the problem; rather, they had hoped to overcome it. Their stubbornness contains many of the elements of classic Greek tragedy in which the mighty are dragged down by their own weaknesses or, in the ironic case of Cooke and Roberts, by their inherent decency. Unlike others who would have simply disregarded Indian treaties and surveyed on the Yellowstone's more promising side, the Northern Pacific decided to test the Yellowstone's western-northern side in order to honor Indian peace treaties.[1]

And where did this decision lead in 1873? First and foremost were Custer's two sharp clashes. If these were minor skirmishes by Civil War standards, they nevertheless received national attention: newspapers sold more copies with Indian warfare headlines than, say, about gigantic coal deposit discoveries. Worse, the fighting was dramatized by Custer's marvelously written, self-serving Official Report, which thrilled schoolboys (and like-minded adults) as much as it frightened Northern Pacific investors. Custer's report piled fears upon an already jittery market. When Cooke's partners (who were spooked the most) shut his company down at 10:30 AM, September 18, the Panic of 1873—the second-worst recession/depression in the nation's history—began.

Logic dictates that had Stanley stayed on the Yellowstone's eastern side, no substantive fighting would have occurred. Quite simply, after the Mandan fiasco, Sitting Bull decided to hunt on the Yellowstone's western side and avoid Stanley. Unlike 1872, few Indian elders were looking for a fight. While Stanley and Custer might have come up the river as far as today's eastern Billings—which Northern Pacific surveyors had reached in 1871 and '72—no *major* clashes would have occurred.[2] With a relatively peaceful Bismarck round-trip, perhaps Cooke could have raised the money to begin 205 miles of construction toward the Yellowstone in 1874. Guarding Northern Pacific workers would have been the army's first priority, and thus Custer would have been on a far tighter leash, although quite likely in command.

And what would Sitting Bull and his allies have done then? Numerous scenarios are available for us armchair strategists. But truthfully, could anyone have imagined a Little Big Horn if there hadn't been one? Or a Pearl Harbor or a 9/11?

As for the factual: with 1873's Panic under way, the Northern Pacific went into bankruptcy within days. To the surprise of many, the railroad

---

[1] Ultimately the Northern Pacific paid the Crow a token $25,000 for hundreds of square miles of land for the Yellowstone's eastern-southern right of way.

[2] One assumes that numerous young warriors—to prove their mettle—would have engaged in hit-and-run attacks, but nothing of the scope of August 11 or even August 4.

survived through good management and the judicious sale and barter of its huge land grants. By 1878 construction was again under way. It picked up steam (a nice railroad idiom) and finished its transcontinental construction in September 1883. Nevertheless construction costs and lavish expenditures outpaced funding. By 1884 the railroad was effectively broke again and within a decade had lost its primacy to the Great Northern. Ultimately the competition between the two meant little; between air and automotive competition, like so many other postwar railroads, in 1970 the Northern Pacific and the Great Northern were both folded into the Burlington Northern Santa Fe system.

Selected Bibliography

Primary Sources

Allen, Joel Asaph. "Notes on the Natural History of Portions of Dakota and Montana, Being the Substance of a Report to the Secretary of War on the Collections Made by the North Pacific Railroad Expedition of 1873, Gen. D. S. Stanley, Commander." *Proceedings of the Boston Society of Natural History* (Boston) 17 (October 1874).

Atwater, Elizabeth Rodgers, ed. "Letters of Montgomery Meigs . . . 1872–1873." Master's thesis, University of Montana, 1937. Microfilm # 140, Montana Historical Society, Helena.

Barrows, Samuel June. *New York Tribune*, June 28, July 3 and 25, August 5, 19, and 23, and September 6, 8, and 9, 1873.

———. "The Northwestern Mule and His Driver." *Atlantic Monthly*, May 1875.

Berry, A. L. *Yellowstone Expedition of 1873, The Narrative of* . . . Alderman Special Collections Library, University of Virginia, Charlottesville, not dated, Rosser Collection, 1171-g-h-j, box 12.

Brace, Charles L. "The New North-West." *New York Times*, July 29, August 5, 6, 8, and 9, 1873.

Braden, Charles. "An Incident on the Yellowstone Expedition of 1873." *Journal of the United States Cavalry Association* 15, no. 54 (October 1904) (unsigned but written by Braden).

———. "The Yellowstone Expedition of 1873." *Journal of the United States Cavalry Association* 16 (October 1905).

Bradley, Luther P. 1873 Diaries and letters. U.S. Army Military History Institute, Carlisle, Pa.

Corliss, Augustus W. *The Yellowstone Expedition of 1873*. Handwritten, May 7, 1896; Indian Wars Miscellaneous Collection folder, U.S. Army Military History Institute, Carlisle, Pa.

Custer, George Armstrong. "Battling with the Sioux on the Yellowstone." *Galaxy Magazine*, 1876. Reprinted in Paul Andrew Hutton, *The Custer Reader*. Lincoln: University of Nebraska Press, 1992.

———. "Official Report of the Tongue River and Big Horn Fights." *New York Tribune*, September 6, 1873. Reprinted in Elizabeth B. Custer, *"Boots and Saddles" or, Life in Dakota with General Custer* (1885), University of Oklahoma Press.

Konopicky, Edward. 1873 letters. Translated by Paul M. Baltay, German edition edited by Robert Pils, 2002. Museum of Natural History, Vienna.

Northern Pacific Railroad. *Clippings and Scrapbooks, 1866–1904, Office of the Secretary of the Northern Pacific.* Minnesota Historical Society, M-522.

———. *Secretary's Department: Unregistered Letters Received and Related Records, Undated and 1864–1876.* Minnesota Historical Society, M-459.

Phelps, William Franklin. "Notes on the Yellowstone." *National Teacher's Monthly* 1, no. 1 (November 1874) through 1, no. 5 (March 1876).

Rosser, Thomas Lafayette. *Letters and Diary for 1873.* Alderman Library, University of Virginia, # 1171-g-h-j, box 1, box 16.

Smalley, Eugene Virgil. *History of the Northern Pacific Railroad.* 1883; reprint, New York: Arno Press, 1975.

Stanley, David Sloan. *Personal Memoirs of Major-General D. S. Stanley, U.S.A.* Cambridge: Harvard University Press, 1917; reprint, Gaithersburg, Md.: Olde Soldier Books, 1987.

Varnum, Charles A. *I, Varnum: The Autobiographical Reminiscences of Custer's Chief of Scouts: Including His Testimony at the Reno Court of Inquiry.* Edited by John Melvin Carroll. Lincoln: University of Nebraska Press, 1982.

Winser, Henry J. *The Great Northwest: A Guide-Book and Itinerary for the Use of Tourists and Travelers over the Lines of the Northern Pacific Railroad, the Oregon Railway and the Navigation Company and the Oregon and California Railroad.* New York: G. P. Putnam's Sons, 1883.

Books

Ambrose, Steven E. *Crazy Horse and Custer: The Parallel Lives of Two American Warriors.* New York: Anchor Books, 1996.

Barrows, Isabel C. *A Sunny Life: The Biography of Samuel June Barrows.* Boston: Little, Brown, 1913.

Bray, Kingsley M. *Crazy Horse: A Lakota Life.* Norman: University of Oklahoma Press, 2006.

Brown, Mark H. *The Plainsmen of the Yellowstone.* New York: G. P. Putnam's Sons, 1961.

Carroll, John Melvin, ed. *Cavalry Bits, One of Seven Volumes: The Yellowstone Expedition of 1873.* Mattituck, N.Y.: J. M. Carroll, 1986.

Darling, Roger. *Custer's Seventh Cavalry Comes to Dakota.* El Segundo: Upton & Sons, 1989.

Englehardt, Carroll. *Gateway to the Northern Plains: Railroads and the Birth of Fargo and Moorhead.* Minneapolis: University of Minnesota Press, 2007.

Frey, Robert Lovell, ed. *Railroads in the Nineteenth Century.* New York: Facts on File, 1988.

Frost, Lawrence A. *Custer's 7th Cav and the Campaign of 1873*. El Segundo, Calif.: Upton & Sons, 1985.

———, ed. *Some Observations on the Yellowstone Expedition of 1873*. Glendale, Calif.: Arthur H. Clark, 1981.

Hammer, Kenneth M. *Men with Custer: Biographies of the 7th Cavalry*. Edited by Ronald H. Nichols. Hardin, Mont.: Custer Battlefield and Historical Association, 1995.

Hanson, Joseph Mills. *The Conquest of the Missouri: Being the Story of the Life and Exploits of Captain Grant Marsh*. 1916; reprint, Mechanicsburg, Pa.: Stackpole Books, 2003.

Heitman, Francis B. *Historical Register and Dictionary of the United States Army . . . , from September 29, 1789, to March 2, 1903*. 1903; reprint, Urbana: University of Illinois Press, 1965.

Hoopes, Lorman L. *This Last West: Miles City, Montana Territory, 1876–1886; The People, the Geography, the Incredible History*. Miles City, Mont.: privately printed, 1990.

Hutton, Paul Andrew. *Phil Sheridan and His Army*. Lincoln: University of Nebraska Press, 1989.

Innis, Ben. *Bloody Knife: Custer's Favorite Scout*. Bismarck: Smokey Water Press, 1994.

Larson, Robert W. *Gall: Lakota War Chief*. Norman: University of Oklahoma Press, 2007.

Lass, William E. *A History of Steamboating on the Upper Missouri River*. Lincoln: University of Nebraska Press, 1962.

———. *Navigating the Missouri: Steamboating on Nature's Highway, 1819–1935*. Norman: Arthur H. Clark, 2007.

Lubetkin, M. John. *Jay Cooke's Gamble: The Northern Pacific Railroad, the Sioux, and the Panic of 1873*. Norman: University of Oklahoma Press, 2006.

Merington, Marguerite, ed. *The Custer Story: The Life and Intimate Letters of General George A. Custer and His Wife, Elizabeth*. 1950; reprint, Lincoln: University of Nebraska Press, 1987.

Oberholtzer, Ellis Paxson. *Jay Cooke, Financier of the Civil War*. 2 vols. 1907; reprint, New York: Augustus M. Kelley, 1968.

Powell, Peter John. *People of the Sacred Mountain: A History of the Northern Cheyenne Chiefs and Warrior Societies, 1830–1879*. 2 vols. San Francisco: Harper and Row, 1981.

Powell, William H. *List of Officers of the Army of the United States from 1799 to 1900*. 1900; reprint, Detroit: Gale Research, 1967.

Powers, Thomas. *The Killing of Crazy Horse*. New York: Alfred A. Knopf, 2010.

Renz, Louis T. *History of the Northern Pacific Railroad*. Fairfield, Wash.: Ye Galleon Press, 1980.

Sajna, Mike. *Crazy Horse: The Life behind the Legend*. New York: John Wiley, 2000.

Sandoz, Mari. *Crazy Horse: The Strange Man of the Oglalas*. 1942; reprint, Lincoln: University of Nebraska Press, 1992.

Stewart, Edgar I. *Custer's Luck*. 1955; reprint, Norman: University of Oklahoma Press, 1980.

Thrapp, Dan L. *Encyclopedia of Frontier Biography*. 3 vols. 1988; reprint, Lincoln: University of Nebraska Press, 1991.

Utley, Robert M. *Cavalier in Buckskin: George Armstrong Custer and the Western Military Frontier.* Norman: University of Oklahoma Press, 1988.

———. *The Lance and the Shield: The Life and Times of Sitting Bull.* New York: Henry Holt, 1993.

Vestel, Stanley. *Sitting Bull: Champion of the Sioux.* 1932. Reprint, Norman: University of Oklahoma Press, 1989.

Welch, James, with Paul Stekler. *Killing Custer: The Battle of Little Bighorn and the Fate of the Plains Indians.* New York: W. W. Norton, 1994.

Wert, Jeffry D. *Custer: The Controversial Life of George Armstrong Custer.* New York: Simon & Schuster, 1996.

Articles

De Land, Charles E. "Basil Clement (Claymore): The Mountain Trappers." *South Dakota Historical Collections* 11 (1922): 245–390.

Gray, John S. "Bloody Knife, Ree Scout for Custer." *Westerners Brand Book [Chicago]* 17, no. 12 (February 1961): 89–96.

———. "Itinerant Frontier Photographers." *Montana*, Spring 1978.

Hammer, Kenneth M. "Railroads and the Frontier Garrisons of Dakota Territory." *North Dakota History* 46, no. 3 (Summer 1979): 24–34.

Hanson, Joseph Mills, "Thomas Lafayette Rosser." *Cavalry Journal*, March–April 1934.

Heski, Thomas M. "Soldiers, Surveyors, Steamboats & Stanley's Stockade." *Research Review: Journal of the Little Big Horn Associates* 13, no. 2 (Summer 1999): 60–76.

Lass, William E. "Steamboats on the Yellowstone." *Montana*, Autumn 1985.

Lubetkin, M. John. "Clash on the Yellowstone: Monday, August 4, 1873." *Research Review: The Journal of the Little Big Horn Associates* 17, no. 2 (Summer 2003).

———. "'Strike Up Garryowen': August 11, 1873: Custer's Second Battle on the Yellowstone." *Research Review: The Journal of the Little Big Horn Associates* 20, no. 2 (Summer 2006).

———. "Thomas L. Rosser and the Yellowstone Surveying Expedition of 1873." *North Dakota History* 70, no. 3 (2003).

Mangum, Neil. "Gall: Sioux Gladiator or White Man's Pawn?" Fifth Annual Symposium, Custer Battlefield & Historical Museum Association, Hardin, Mont., 1991.

Rolston, Alan. "The Yellowstone Expedition of 1873: Military-Civilian Contentions Regarding Western Exploration and Indian Policy." *Montana*, Spring 1970.

Selmeier, Lewis W. "First Camera on the Yellowstone." *Montana*, Summer 1972.

Stewart, Edgar I. "A Psychoanalytic Approach to Custer." *Montana*, Summer 1971.

Taylor, Joseph Henry. "Bloody Knife and Gall." *North Dakota Historical Quarterly* 4, no. 3 (April 1930).

Acknowledgments

I am most grateful to two college friends: Paul M. Baltay, of Arlington, Virginia, who discovered and translated Edward Konopicky's letters, and Lewis T. Buckman, C.E., of East Berne, New York, who did his best to explain some of the mysteries of surveying and civil engineering. Dave Eckroth of Billings, Montana, generously shared his discoveries and insights with me, and perhaps by the time this is published, he will have found—where so many others have failed—the elusive August 4 battle site. Additionally Jan Taylor of Missoula, Montana, spent many hours transcribing Samuel Barrows's articles, without which this book could not have been written.

William E. Coffee of Billings, Montana, and his sister, Caren L. Coffee, went out of their way to make sure that the reproduction of the late James K. Ralston's magnificent painting of Custer and the 1873 surveying expedition was technically suitable for use in this book. While the painting's size made its use on the cover unfeasible, this is the first that it has been reproduced in print, and this writer could not be more appreciative of the Coffees' efforts.

During the recent recession, no category of public service seems to have taken a greater financial hit than public libraries and state historical societies. Yet despite the cultural shortsightedness of both red and blue elected officials and often paranoid post 9/11 "guardians," these institutions have survived and, once inside, the researcher can find trained personnel who couldn't be friendlier.

For this and *Jay Cooke's Gamble,* help was received from almost two hundred people in various libraries, especially the Fairfax and Arlington County, Virginia, public library systems, as well as the Alderman Special Collections Library, University of Virginia-Charlottesville;

Arlington County; Burlingame Special Collections Library, Montana State University–Bozeman; Harvard University's Houghton Reading Room, Cambridge; the Library of Congress and National Archives in Washington, D.C., and Silver Spring, Maryland; Minnesota State Historic Society, St. Paul; Montana State Historic Society, Helena; North Dakota State Historic Society, Bismarck; South Dakota State Historic Society, Pierre; the U.S. Army Military Academy, Archives, and Special Collections Department, West Point; and the U.S. Army Military History Institute, Carlisle, Pennsylvania.

Invaluable support has also been provided by many members of the Custer Battlefield Historical and Museum Association, the Little Big Horn Associates, and the Northern Pacific Railway Historical Association. Each organization holds annual conferences, publishes high-quality magazines, and is composed of innumerable intelligent, well informed, and friendly individuals who have gone out of their way to be helpful.

Special note should be taken of the late Dr. Lawrence A. Frost's Herculean efforts in documenting 1873's events. Whatever professional disagreements I have with his depiction of David Stanley, without his well-documented research trail, this book would not have been possible.

As a reader often frustrated by good books with bad maps—if any—I have tried to use period or easy-to-read maps. For the latter I am particularly appreciative of the work of Vicki Trego Hill of El Paso, Texas, who took period maps and my rough sketches and turned them into something first-rate.

Hats off to my son, William Andrew Lubetkin, who has painstakingly introduced me to the electronic wonders of the 1990s—perhaps even the twenty-first century. And I especially wish to thank my wife, Linda, a retired high school English teacher, who spent hundreds of hours reading and making critical suggestions and comments. Without her skills, patience, and good humor, this book couldn't have been completed.

Index

Page references in italic type refer to illustrations.

Adair, Lt. Louis Dent, 29
Agassiz, Louis, 36
Alcoholism: David Stanley drunk, 177; David Stanley's reputation for, 16; delays caused by, 33–34; Thomas Rosser and, 151
Alkaline water, 58, 168, 204
Allen, Joel Asaph, *274*; areas of expertise, 65 & n39; Konopicky's comments about, 293–94; Missouri skylark discovery, 132–33; in scientific party, 84
Ammonites, 296 & n3
Ammunition: Indian use of, 27; restrictions on use of, 124; uselessness of, 127; as wagon train cargo, 75 & n1
Amorpha canadenis, 124
Anaxagoras, 122 & n7
Andrews Creek, Dak. Terr., 84
Antelope, 122–24, 147, 205
Antietam, Battle of, 31
Apache Indians, 121n6
Apple Creek, Dak. Terr., 50
Appomattox (campaign), 31, 43
Arapaho Indians, 24, 26
Argillaceous sand, 164
Arikara Indians, 97
Arkansas Flycatcher, 125
Arlington Cemetery, 42
Army and Navy Journal, The, 303, 311–12

Army snafu, 151
Arroyo, 201
Arson, 189
Artemisia, 133
Artillery, 96, 99, 255–56, 263
Ashbend, 300
Ash trees, 171, 199, 295
Astragelus carjocarpus, 124
A-tents, 161 & n7
Atlanta Campaign (1864), 33 & n36
Austin, Gov. Horace (Minn.), 37
Axeman, 52, 90, 93

Babb, Loomis, 93
Bad Lands, 163–70, 174, *281*. See also Mauvaises Terres
Bad Route Creek, 200 & n16, 297
Baker, Capt. Edward, 82, 191
Baker, Maj. Eugene M., 24–27
Baker's Trail, 294–95
Baliran, Augustus (sutler), 234 & n24
Ball, Pvt. John, 232 & n21, 250
Band, Seventh Calvary, 118, 254 & n14
Barrows, Samuel, articles by: Bad Lands description, 163–70; Battle of the Big Horn, 240–47, 250–51, 254–56; buffalo and arrival at Yellowstone, 187–91; Dakota mud and hailstorms, 126–33; Indian attacks, 221–33;

Barrows, Samuel, articles by (*continued*): life on the expedition, 116–20; logistics of the Yellowstone Survey, 99–105; organization of the Yellowstone Survey, 77–85, 87–88; return home of the expedition, 302–303; river obstacles, 153–58; scenery and campsites, 158–63; steamboats on the Yellowstone River, 192–94; success of the Yellowstone Survey, 301–302; weather and scientific observations, 120–26
Barrows, Samuel June, 34–37, 75, 142, 183, *266*
Battalion, 104–105, 111
Battle of Britain, 185
Battle of Poker Flats, 26
Battle of the Big Horn, 240–56
Bears, grizzly, 195–96, 258
Beaver Creek, 171 & n17, 174n22, 176n28
Beecher, Henry Ward, 35
Beecher's Island, 188n2
Beef cattle, 96
Beethoven, Ludwig von, 148
Bell flower, 163
Bennett, Prof. Charles W., 65n39, 85, 294
Benteen, Capt. Frederick, 165 & n12, 185n1
Berry, A. L., 239, 262–63, *269*
Big Hill, *215*, 297 & n4
Big Horn Canyon, 174 & n22
Big Horn River, Mont. Terr., 33 & n36, 244 & n7
Big Knife River, Dak. Terr., 141, 162
Big Mound, battle of, 20
Big Muddy Creek, Dak. Terr.: attempts to cross, 155–58; bridge washed out, 153–54; military arguments and, 140n13; as an obstacle, 153–55
Big Porcupine Creek, Mont. Territory, 111n8
Billings, Mont., 313
Binge drinking. *See* Alcoholism
Bird eggs, 132–33
Bismarck, N.Dak.: description of, 72–73, 78 & n3, 206–207; growth of, 71; nightlife in, 51

Bismarck Tribune, 142n15, 207, 303–307
Bison. *See* Buffalo
Bitter Creek, Wyo. Terr., 200 & n15
Bitters, 180
Blackfeet Sioux, 26, 62, 233
Black Hills Expedition (1874), 36, 186, 313
Black Kettle, 237n1
Blacksmiths, 110, 149 & n20
Blacktail deer, 174
Bloody Knife, *275*; background, 224 & n9; in battle, 228; skills of, 241, 246
Blue Danube waltz, 148
Bluffs, 133, 193–94, 198–99
Board of Officers, 299
Bonds, federal, 17, 29
Boston, Mass., 66, 95
Bouyer, Mitch, 27
Bozeman Pass, 23, 26
Bozeman Trail, 20, 174n22
Brace, Charles Loring, 186, 204–207, *270*
Brackett, George A., 60 & n25
Braden, Lt. Charles, 251 & n12, 306
Bradley, Lt. Col. Luther P., *268*; abandons stores, 299; arrival at Fort Rice, 41, 62; background, 43–44; diary of, 60–63, 105, 106, 170–72, 139–42, 200–201, 234–35, 259–60, 294–301
Bradley, Lt. Col. Luther P., letters: arrival at Grand River Agency and the 7th Calvary, 63–65; Bad Lands and Yellowstone scenery, 173–75; delays and mosquitoes, 105–106; expedition progress and hailstorms, 142–45; Sioux attacks, 258–59
Brady, Matthew, 67n41
Brainerd, Minn., 18, 23
Breckinridge, Minn., 301 & n8
Brewer's Blackbird, 125
Bridges, 50, 130–31, 188
Brogans, 28
Brown coal, 164 & n11
Brulé Sioux, 311
Brush, Lt. Daniel H., 104 & n4, 117, *273*
Buffalo: causes of disappearance, 188–89; corpse remains, 68; decline of, 20; killing of, 257
Buffalo grass, 164

Buffalo Island, 62
Bull-boats, 246 & n11
"Bull without Hair" (Minneconjou), 311
Bunting (bird), 125
Bunyan, John, 129 & n9
Burial ceremony, 175
Burleigh, Dr. Walter, 91 & n18
Burlington Northern–Santa Fe Railroad, 315
Buttes, 125, 158–60, 168–69

Cabin Creek, Mont. Terr., 173, *184*, 201
Calhoun, Fred, 180n37
Calhoun, Lt. James, 180n37, 229, 238
California, 17–18, 166, 186
Cambridge, Mass., 69, 293
Campbell, Lt. Lafayette E., 61 & n28
Camp Canby, 151 & n1, 195
Camp Ferry, 64
Camp Gronewald, 304 & n12
Camp Mud Creek, 144
Campsites, 26, 237, *287*
Camp Terry, 62
Cannibalism, 97
Canadian Pacific Railroad, 18
Canby, Gen. Edward, 151 & n1
Canoe, 67 & n42
Capitol Hotel, Bismarck, 207
Carbines, 110–11, 227, 254
Carlin, Lt. Col. William P., 68 & n43
Carlisle, Pa., 31
Carte de visite, 161 & n8
Cass, Gen. George Washington, 29, 42, 270
Cattle, 75, 118–19, 179 & n35
Cavalry, 16, 115, 262
Cedar Creek, battle of, 176 & n29
Central Park, New York City, 66, 306
Chainman, 90, 93
Chapin, Isabel Hayes (Mrs. Samuel June Barrows), 35
Chattanooga, battle of, 202n18
Cheyenne Agency, 62 & n31
Chicago, Ill., 49, 66, 203
Chicago Symphony Orchestra, 118 & n3
Chickamauga, battle of, 299 & n6
Children's Aid Society, 186

Chummey, 73 & n50
Churchill, Sir Winston, 185
Civil War, 17, 35
Clark, Capt. William (Lewis and Clark), 192
Clarke, Capt. Charles E., 57 & n23
Clement, Antoine, 311 & n21
Clement, Basil (Clemmo), 117 & n2, 131, 187, 311n21
Clifford, Louis, 243
Clouds, 159
Coal, 206, 295, 307
Cocktails, Glendive, 179
Coffee, 91, 140n13, 245
Colfax, Schuyler (vice president), 203 & n19
Colorado, 20, 170
Column, compact, 167
Commissary stores: chief commissary officer, 155, 158; excess of, 191; required amount, 163; wagon load of, 119
Condensed milk, 120
Confederate army: Appomattox and, 31; David Stanley in, 33 & n35; interview with George Custer and Thomas Rosser, 177; Thomas Rosser and, 43; W. Milnor Roberts and, 42
Congregational church, Bismarck, 207
Congress, U.S., 17, 81
Continental Divide, 18–19
Cook, for surveyors, 93
Cook, Rev. W. W., 61
Cooke, Jay: bank of, 29; financial problems and failure of, 23, 307–309; construction route, 313–14; sense of purpose, 18–19
Cookery, Jay, 222
Corliss, Capt. Augustus W., *271*; background, 97–98; report on the Yellowstone Survey, 108–12
Corral, 110
Cottonwood trees, 199, 225, 241
Coulson's Island, Mont. Terr., 299–300
Council Bluffs, Iowa, 17n6
Crazy Horse, 220 & n2
Crazy Mountains, 26
Crittenden, Col. Thomas L., 68 & n44

Crittenden's Island, Mont. Terr., 145 & n18
Crofton, Maj. Robert E. A., 172, *274*; Eban Crosby's death and, 28; fear of Indians, 16; personality, 191 & n5
Crosby, Lt. Eban, 28
Crow, Private J. R., 229
Crowell, Lt. William H. H., 195 & n9
Crow Indians, reservation of, 23, 42n4
Cultivation practices, 205
Curlew (bird), 125, 204
Custer, Lt. Col. George Armstrong, *265*; arrest of, 179–81; August 4 fighting, 219–35, 297; August 11 fighting, 237–63; background, 30–32; beef cattle on pursuit, 235, 237n1; biographies of, 30n30; *Bismarck Tribune* interview, 303–306; friendship with W. Milnor Roberts, 42; Indian villages and, 237n1; interview with, 177–78; on Musselshell River, 296; official report, 314; physical description, 178; post-expedition analysis, 314; reputation, 15–16; return to Bismarck, 298, 301–302; uniforms worn, 304–305
Custer, Lt. Thomas W.: Indian attacks and, 219–20; military recognition, 223n7; in scouting party, 223–24
Custer's bridge, 153–54, 153n3
Custer's Creek, Mont. Terr., 200 & n17

Dahna, A. O., 93
Daily line of march, 103
Daily schedule during expedition, 103–104
Dakota Division, 72, 178
Dakota mud, 127–31
Dakota Territory, 76, 77, 99
Dansville, N.Y., 35
Danube River, 182
Darwin, Charles, 30
Dashiell, Henry, 142n15
David Copperfield (Dickens), 154 & n4
Davis, Henry C., 167n14
Davis Creek, Dak. Terr., 167 & n14
Davis Island, N.Y., 155 & n5
Dead Buffalo Lake, battle of, 20

Declaration of Independence, 186
Decoys, 220n5, 225–26
Deer, Blacktail, 174
Deerfield, E. P., 93
Deerkill Mountain, battle of, 165 & n13
Deflection angle, 90
Department stores, 79n4
Depression, 16, 18, 314
DeRudio, Lt. Charles C., 87 & n12, 97
Desert, 148, 166
Desk engineers, 155
De Smet, Father Pierre-Jean, 312 & n23
De Smet (steamboat), 62–63
Dewy, Louise Ione, 63n34
Dickens, Charles, 154n4
Digitalis, 124
Division of the Missouri, 42n1
Doane, Lt. Gustavus C., 46 & n
Dogs, 122–23, 137, 256. *See also* Given (dog)
Dog-Teeth Buttes, 125–26
Dog tents, 304 & n13
Dough boys (infantrymen), 255 & n16, 303
Dougherty, Lt. William E., 128 & n8
Dried-up streams, 168
Drought, 189
Drowning in Missouri River, 70
Duluth, Minn., 19 & n10, 84, 206, 313
Dust, 99, 168, 296

Eagle's Beak, *283*
Eckelson, Albert O., 50 & n11, 176, 302, 309–11
Edwinton, N.Dak., 55–58. *See also* Bismarck, N.Dak.
Eggs, bird, 132–33
Eighth Infantry, 98, 306
Eleven-Mile Creek campsite, 119
Elk, 195, 205, 259
Elk Point, S.Dak., 77
Elliott, Maj. Joel H., 32
Emigration, 205–206
Emissaries, 24
Engineer corps: army opinions of, 155; during Indian attacks, 263; tasks of, 89–90

INDEX

Engineering. *See* Surveying, description of
Engineering party, 49, 92–93
Equipment carried, 82

Fagin. *See* Phelps, William F.
Fall River, 66
Falstaff's ragamuffins, 28
Fargo, N.Dak., 20
Far West (steamboat), 145–46, 145n16, 190
Fat, pork, 79
Federal bonds, 17, 29
Feldspathic rock, 125
Ferries on Yellowstone River, 46
Finger Lakes, N.Y., 35
Fires, signal, 296
Fish, Hamilton (secretary of state), 35
Five Corners, New York City, 186
Five Forks, battle of, 177
Flagmen, 90, 92
Flatcar, 53
Fog, 165
Forage, 81, 156, 243
Forrest, Gen. Nathan Bedford, 30n29
Forsyth, Lt. Col. James W., 188n2
Forsyth, Maj. George A., 188 & n2
Fort Abercrombie, Dak. Terr., 75 & n2
Fort Abraham Lincoln, Dak. Terr.: carpenters going to, 204; Indian attacks at, 72–73; location, 79; route to Powder River, 47
Fort Bedford, 190
Fort Benton, Mont., 71 & n47, 246, 291
Fort Breckinridge, 77
Fort Buford, Dak. Terr.: arrival at, 197; distance to, 195, 300; ferries at, 46; supplies and, 188, 190
Fort Ellis, Mont. Terr., 23, 46
Fort Keogh, Mont. Terr., 234 & n26
Fort Laramie, Wyo., 312n23
Fort McKean, Dak. Terr. (Fort Abraham Lincoln), 58
Fort Randall, Dak. Terr., 61 & n27
Fort Rice, Dak. Terr.: activity at, 73–74; arrival of Luther Bradley, 62; description of, 99, population at, 51; winter at, 306

Fort Smith, Mont. Terr., 174n22
Fort Snelling, Minn., 30, 301
Fort Stevenson, Dak. Terr., 300
Fort Sully, Dak. Terr., 24, 33
Fort Union, Dak. Terr., 25
Fort Wadsworth, Dak. Terr., 207
Fossils: marine, 296; prospect of finding, 30; sea of, 112; of vegetables, 174; of wood, 164
Foundry, 66
Fourth of July, 162
Foxglove, 124
Freedman's Hospital, 35
Freight, 72, 146, 303
Frémont, John L., 232 & n20
French, Capt. Thomas, 230 & n19
Friendly fire, 124, 147
Frontiersman, 18
Frost, R. G., 243
Fry, A., 93
Fuel, 206

Galaxy Magazine, 229n15
Gale, 196, 296
Gall, 27–29, 43, *253–54*, 277
Gamblers, 73n50, 205, 206
Garden of Eden, 100
Garden of the Gods, Colo., 170
Gardner, Alexander, 67n41
Garry Owen, played, 254
Gastropods, 160
Gatling gun, 27, 28
Geese, 194
Geography, stratigraphical, 160
Geologists, 84, 174, 182
Gettysburg, Battle of, 30, 177 & n32
Geysers, 206
Given (dog), 122, 162
Glacial drift, 125
Glendive cocktails, 179
Glendive Creek, Mont. Terr.: bluffs at, 309; construction to, 313; crossing of, 171 & n18; steamboats from, 172; supply shipments and, 75
Glendive Falls, Mont. Terr., 299 & n5
Gold strikes, 20
Graham, Charles, 273

Grain elevator, 206
Grand River Indian Agency, Dak. Terr.:
    location, 56 & n22; Luther Bradley at,
    62 & n32; trade with Indians at, 63
Grant, Julia Dent (First Lady), 29 & n25
Grant, Lt. Col. Frederick Dent (son),
    272; departure of, 194–97; description
    of, 53 & n16; reason for joining the
    survey, 29
Grant, President Ulysses S., 18
Grapeshot, 256 & n17
Grass, buffalo, 164
Gravel, 145, 199
Graves, of Indians, 95, 221, 240
Grease, wagon, 299
Great Chicago Fire, 66
Great Northern Railroad, 315
Greek tragedy, 313–14
Gregory, Rear Adm. Francis H., 61 &
    n29
Grizzly bear, 195–96, 258
Gulf of Mexico, 170
Gullies, 119
Gunter's Chain, 90n16
Gypsum, 164

Hailstone Creek, Dak. Terr., *279*
Hailstorms, 131–32, 136–37, 144
Hale, Capt. Owen, 222 & n4, 230, 245–55
Hancock, Winfield Scott, 31
"Happy Hunting Grounds," 232
Hard tack, 118, 158, 245
Harris, Stephan, 28
Hart, Capt. Verling K., 180 & n36
Harte, Bret, 26
Harvard Divinity School, 36–37, 237
Hayden, Ferdinand Y., 48n10
Haydon, John A., 48 & n
Hazen, Col. William B., 202 & n18, 300,
    303
Headquarters Hotel, Fargo, 54 & n18
Heart River, Dak. Terr., 140, 143, *273*
Heart's Butte, Dak. Terr., 74, 125
Heat, 94, 121–22, 148, 182
Henry rifles, 231, 240
Herbarium, 133

Hoe, Richard M., 34–35
Hoffman, Private, 70
Hogs, 79, 190n3
Home-seekers, 308n17
Homesickness, 134, 176, 182
Honsinger, Dr. John, 222–23, 230–32
Hoop Pole Country, 146
Hoosac Railroad Tunnel, Mass., 120 &
    n5
Horses, 96, 99, 102, 246, 247
Hotel Sherman, 66
Hudson Palisades, 194
Humidity, 16
Hunkpapa Sioux, 15 & n1, 27–29. *See also*
    Unepapa Indians
Hunters, 82–83, 124, 244–46
Hunting grounds, Indian, 15
Hydraulic energy, 167
Hydraulic mining, 166
Hypochondriac, 43

Ice, 21, 55, 155. *See also* Hailstorms
Idaho, 18, 20
Ida Stockdale (steamboat), 54–58, 54 & n17,
    54n20, 92
Impedimenta, 82, 101
India, 35
Indian attacks, 27–29, 219–20, 225–35,
    237–39, 247–59
Indians: Apache, 12n6; Arapaho, 24,
    26; Blackfeet Sioux, 26, 62, 233; Black
    Kettle, 237n1; Bloody Knife, 224 & n9,
    241; Brulé, 311; "Bull without Hair,"
    311; campsites of, 160–61; Cheyenne
    Agency, 62 & n31; Crazy Horse, 220
    & n2; Crow, 23, 42n4; Gall, 27–29, 43,
    277; and grave robbery, 202; Hunkpapa
    Sioux, 15 & n1, 27–29; hunting
    grounds, 15; Lakota Sioux, 20, 220;
    Little White Swan, 311 & n20; lodges,
    20, 160–61, 240 & n4, 241 & n6;
    Minneconjou, 219, 311 & n20; Modoc,
    15n1; Mr. Lo, 198 & n13; Nez Perce,
    15n3, 219; "Ogalalla" (Oglala), 233 &
    n23, 312; Pawnee, 312; Plains, 312n23;
    "Rain-in-the-Face," *272*; recovery of

INDEX 329

casualties, 257; Red Cloud, 20; Red Ears, 311; "Ree," 80; Sioux, 20; Spotted Eagle, 24; Unepapa, 233; villages abandoned, 235, 237n1, 240–42. *See also* Sitting Bull
Infantry, 27–28, 74, 103–104, 255
Infrastructure, 19
Inman, Capt. Henry, 171 & n17
Inman Creek, Mont. Terr., 171n17
Inman's Fork, Mont. Terr., 176 & n28
Iron, 125
Iron road. *See* Railroads
Irrigation, 199–200

Jamaica ginger, 180
Jamestown, N.Dak., 53, 72, 204
Jay Cooke & Co., 18, 292, 307–309
Jewell's Chute, Mont. Terr., 146
Jones, Lt. James H., 82, 251 & n13
Jordan, Edward C., 92n20, 267
Josephine (steamboat), 145n17, 201, 298–302, *284*
Journal of the United States Calvary Association, 230n18
Judith Mountains, 258

Kansas frontier, 32
Katie Kountz (steamboat), 69–70, 69n45
Kearney, Gen. Phil, 31
Kegs, 156–58
Kellogg, Mark, 71–73
Ketchum, Lt. Hiram H., 82, 105 & n5
Key West (steamboat): arrival of, 172, 173, 190; cost of renting, 192; description of, 145n16, 193; expected arrival of, 51; ferrying, 172; Frederick Grant and, 195; hiring of, 145; problems with, 145–46; store delivery by, 172
Kimball, Assistant Surgeon J. P., 82
Kingbird, 124–25
Konopicky, Edward M., 44; education, 36–37; pages missing from letters, 239n25; in scientific party, 84–85
Konopicky, Edward M., letters of: arrival at Yellowstone River, 181–83; to Dr. Franz Steindachner, 293–94; Frederick Grant and rattlesnake, 181; on Indian attacks, 256–58; Indian encounters, 233–34; life on the expedition, 147–49; preparation for the survey, 94–95; travel to Fort Rice, 65–69
Kountz, William J., 69n45

Lake Erie, 19
Lake Michigan, 66
Lake Superior, 48, 84, 291, 313
Lakota Sioux, 20, 220
Land grants, 17
Land speculation, 21
Laramie plains, 143
Lazelle, Capt. Henry M.: assignment, 64 & n36; description of, 85; visit with Edward Konopicky, 148–49, 182
Leveler, 90, 92
Lewis, Capt. Meriwether, 192
Lightning, 120–21, 144, 147, 196
Lignite, 164 & n11, 194
Linsley, Daniel, 50 & n12
Lithographic rotary printing press, 34
Little Big Horn: Charles DeRudio and, 87n12; George Armstrong Custer and, 30; Long Dog and, 311n22; speculation on, 186
Little Missouri River: arrival at, 171, 173; Bad Lands and, 163; crossing of, 47–48; in Eckelson's report, 311; Rosser and, 165; streams and, 170
Little Porcupine Creek, 235, 296
Little White Swan, 311 & n20
Lo, Mr., 198 & n13, 221n2, 262
Lodges, 20, 160–61, 240 & n4, 241 & n6
London University, 243
Long Dog, 311–12, 311n22
Long Island Sound, 155
Longstreet, Gen. James, 32
Loring, D., 93
Lorris, Col., 63
Lounsberry, Clement, 30, 303–306, 303 & n11

Mails, 137, 201, 298
Makoshika State Park, 174n23

Man-Who-Goes-in-the-Middle (Gall), 27–29, 43, *277*
Maple River, Dak. Terr., 205
Marches, daily, 111–12. *See also under* Bradley, Lt. Col. Luther P.
Maria's Creek Massacre (1870), 26, 237n1
Marine fossils, 296
Marsh, Grant (steamboat captain), 92 & n19, 190, 299
Marsh, Othniel C., 174 & n24
Mauvaises Terres, 167, 190. *See also* Bad Lands
May Lowry (steamboat), 304
McClellan, Gen. George B., 31
McCown's bunting (bird), 125
McPherson, General James B., 33
Mead, G. W., 51 & n15, 205
Meadowlark, 125
Mechanics, 132
Medicine, 161, 312
Medicine man, 228, 232
Medora, N.Dak., 23, 167n14, 175n27
Meigs, Lt. John, 42–43
Meigs, Montgomery, *268*; background, 37; education and family, 42–43; *Ida Stockdale* and, 41; and Rosser, 176
Meigs, Montgomery, letters of: hailstorm, 144; Indian attacks, 261–62; life on the expedition, 197–200; real estate and *Ida Stockdale*, 52–58
Meigs, Montgomery (Sr.), 42
Melstone, Mont., 295n2
Metropolitan Hotel, 59
Mexican War, 16–17, 30n29
Micawber, Wilkins, 154 & n4
Miles City, Mont., 234 & n26, 297n4, 313
Milk, condensed, 120
Miner (steamboat), 62 & n33, 91
Mineral deposits, 19
Mining, hydraulic, 166
Minneapolis, Minn., 59n24, 204, 239
Minneconjou Sioux, 219, 311 & n20
Minnesota, 17, 18
Minnesota River, 301
Mississippi River, 301
Mississippi Road, 84
Missouri River, *210–11*; channels of, 56–57; drinking from, 35, 95; drowning in, 70; Little Missouri River, 47–48; map of, *22*; march to Yellowstone, *172*; scenery near, *209*; to Yellowstone River route map, *86*
Missouri skylark, 132–33, 164
Moccasins, 241
Modoc Indians, 151n1
Molesworth, Louis, 243
Montana Territory, 233, 256
Montauverti, 198
Moorhead, Minn., 18, 23
Mormons, 100
Morse code, 35
Mosquitoes, 67, 120, 143
Mountain sheep, 174, 198, 294
Moylan, Capt. Miles, 223 & n6
Mud, 115, 119–20, 127–31
Mud Creek and campsite, 138, 144
Mule Creek, 139
Mules, 96, 102, 110, 157, 174, 244–46
Mulloy, P., 93
Mur de Slace, 198 & n14
Musselshell River, Mont. Terr., *290*; arrival at, 257–58; description of, 294–95, 310; as destination, 182 & n39
Musselshell Valley, 199
Myers Bridge, Mont., 235n28

Napoleon artillery piece, 27
Napoleon III, 87n12
National Academy of Science, 85
National Road, 42n3
Native Americans. *See* Indians
Nebraska, 17, 18
Nettre, Prof. Lionel R., 84, 233, 257
Nevada, 17, 18
"New North-West" (Brace), 186
New Rochelle, N.Y., 155n5
New York City, 21, 34, 155n5
New York Herald, 70–71, 142
New York Times, 88–91, 204–207, 304–305
New York Tribune: Bad Lands description, 163–70; Battle of the Little Bighorn, 240–47, 250–51, 254–56; buffalo and arrival at Yellowstone, 187–91; Dakota mud and hailstorms,

126–33; Indian attacks, 221–33; life on the expedition, 116–20; logistics of the Yellowstone Survey, 99–105; North Dakota map, *76*; organization of the Yellowstone Survey, 77–85, 87–88; return home of the expedition, 302–303; river obstacles, 153–58; scenery and campsites, 158–63; steamboats on the Yellowstone River, 192–94; success of the Yellowstone Survey, 301–302; weather and scientific observations, 120–26
Nez Perce Indians, 15n3, 219
Ninth Infantry, 98, 306
Normal school movement, 152
Norris, Gillman, 244 & n8, 291
North Dakota, *76*
Northern Pacific Railroad: board of directors meets Rosser, 203; construction of, 42 & n4, *278*, 313–15; financial problems, 29, 292; formation of, 16–19; German fund-raising and, 78 & n3; Indian difficulties, 291–92; investors, 314; mismanagement of, 20–23; planned route, 18; route omissions, 301n8
Northern Pacific Railroad Expedition. *See* Yellowstone Surveying Expedition (1871); Yellowstone Surveying Expedition (1872); Yellowstone surveying expeditions (1873)
Norway pine, 262
Nussdorf, Austria, 182 & n38

Oats, 164
O'Fallon Creek, Mont. Terr., 27, 202
Oglala Sioux, 233 & n23, 312
Old Dominion, 177 & n32
Omaha, Nebr., 205, 298, 303, 306
One Law in Nature (Lazelle), 85 & n11
Ord, Brig. Gen. Edward O. C., 64 & n38
Orderly, 226, 242, 250
Order of calls, 103–104
Oregon, 18
"Outcast of Poker Flats, The" (Harte), 26

Pacific Railroad Acts (1862, 1864), 17

Pacifism, 42
Paint, war, 241
Paleontologist, 174 & n24
Panic of 1873, 18, 292, 314–15
Passion in Tatters (Thomas), 138 & n12
Paulin, 156
Pawnee Indians, 312
Peninah (steamboat), 146–47, 190 & n4
Petrification. *See* Fossils
Petrified wood, 164
Phelps, William F.: background, 37; illustration of, *271*; interview with George Custer and Thomas Rosser, 177–78; mention by Roberts, 93; mention by Rosser, 107; physical description of, 152
Philadelphia Evening Bulletin, 180–81, 308–309
Picket guard, 110, 112
Picket rope, 157
Pickles, 299
Pile bridges, 50
Pine, Norway, 262
Pioneers, army, 119 & n4, 128–30, 189
Pittsburgh, Penn., 66
Pizi (Gall), 27–29, 43, *277*
Plains Indians, 312n23
Plenty Lice, 26
Pole houses, 72
Pompey's Pillar, *209*; description of, 261–62; distance to, 310; expedition reaches, 258, *285*; size estimates, 260 & n20
Ponies, 63, 229, 241
Pontoons, 154
Porcupines, 306
Pork fat, 79
Postal service, 137, 148, 201, 298
Powder River, 47–48, 163
Prairie, 67–68, 138, 161, 204
Prairie dogs, 68, 181, 187
Prairie fire, 23, 28, 127
Prairie Hotel, Fargo, 67
Printing press, lithographic rotary, 34
Profile paper, 90
Prostitutes (soiled doves), 51, 71
Provisions, 95

Pryor's River, Mont. Terr., 84 & n10
Puget Sound, 84, 291, 313
Pup tents, 304n13
Purple lupine, 124
Pywell, William R.: background, 67 & n41; misplaced equipment, 126; practices sleeping, 182; in scientific party, 84; survey supplies of, 87

Quaker, 42
Quarantine, 62–63, 64n35
Quartermaster's Department, 85, 154
Quartzite, 125, 133
Quicksand, 171

Rabbits, 204
Raft on Yellowstone River, 246
Railroad Creek, Dak. Terr., 175n27
Railroads: bring civilization, 160; Burlington Northern–Santa Fe Railroad, 315; Canadian Pacific Railroad, 18; construction of, 88–91, 314–15; Great Northern Railroad, 315; Pacific Railroad Acts, 17; Rutland Railroad, 21; transcontinental, 16–18, 315; trial lines for, 89–90; Vermont Central Railroad, 21. *See also* Northern Pacific Railroad
Rain, 71–72, 115, 120–22, 131–32
Rainbow, 159
Rain-in-the-Face, *272*, 311n22
Ralston, John K., 213
Rapids, 46, 197, 299 & n5
Rations: Indians and, 311–12; issuance of, 104–105; lack of, 118, 255; necessary amount of, 101
Rattlesnakes, 181
Ray, Lt. Phillip H., 82, 121 & n6, 155–58
Real estate speculation, 52–53
Rear guard, 139–41, 167, 295–97
Recession, 16, 18, 314
Red Cloud, 20
Red Ears, 311
Red River Valley, 19, 77–78, 205
Red tape, 145, 154
"Ree" Indians, 80, 97, 246
Reinforcements, 167, 238, 250

Religious services, 73
Reunion, surveyors, 30n27, 268
Reveille, 116
Revolver, 254, 257
Reynolds, Charley, 82–83, 83n6, 140, 244, 291
Rifles, 108, 165, 240
Roberts, Annie (Mrs. G. W. Yates), 42
Roberts, W. Milnor, *269*; background, 42; as chief engineer, 23; and Jay Cooke & Co. collapse, 308 & n18; letters detailing survey preparation, 91–94; letters of survey instructions, 47–52; and railroad route location, 313–14
Rock Springs, Wyo., 200n15
Rocky Mountains, 167, 192
Rodman rifled artillery pieces, 75 & n1, 96. *See also* Artillery
Rodsman, 90, 93
Rope, picket, 157
Rosebud River, 182, *184*, 310
Roses, wild, 164
Rosser, Lizzy (Winston), 59 & n24, 92n21, 203
Rosser, Thomas L., *267*; articles by, 108; background, 43; interview with, 177–78; as lead surveyor, 27, 108, 143; leaves survey, 199, 202; letters of, 107, 136–39; physical description, 178; pursuit by Gall, 28–29; report from Albert Eckelson, 309–11; success of, 307; survey preparation and, 46, 50, 51–52
Rosser, Thomas L., diary and letters of: Custer-Stanley reconciliation, 175; fears Custer will arrest Stanley, 177; July 8 to July 23, 175–77; July 24 to August 19, 201–204; June 13 to June 19, 106, 108; June 20 to July 7, 133–36; May 22 to June 12, 58–60; Stanley drunk, 177
Russell Indian Agency, 61
Rutland Railroad, 21
Rye, 162, 164

Sage, 99
Sand, argillaceous, 164
Sandbars, 56, 58

INDEX   333

Sand Creek massacre (1864), 170, 237n1
Sandstones, 125
Saskatchewan region, 205
Sauerkraut, 299
Scandinavian settlers, 18
Schubert, Franz, 148
Scientific observations, 124–26, 132–33, 163–65
Scientific party, 84–85, 87, 303. *See also* Letters of Konopicky, Edward M., letters of
Scouts: bull-boats and, 238; compensation for, 82; as guides, 117–18; hunting duties of, 68; Indian attacks and, 231; Ree Indians as, 80. *See also* Bloody Knife
Seashells, 296n3
Sea turtles, 296
Sentinel Buttes, Dak. Terr., 171, 175–76
Seventeenth Infantry, 95–96
Seventh Cavalry: arrival of, 41, 302; assignment of, 30; band, 118, 139, 254; in combat, 247–55; delay march, 140; organization of, 96; review of, 102–103
Seward, William H. (secretary of state), 35, 152
Shacks (pole houses), 72
Shale, 133, 168–69
Sharpe rifles, 165
Sheep, mountain, 174, 198, 294
Shelter tents, 161 & n7
Shenandoah Valley, 177n32
Sheridan, Lt. Gen. Philip H.: Frederick Grant and, 29; letters regarding survey preparations, 45–47; meets with Rosser, 203; Northern Pacific and, 24
Sherman, Gen. William T., 17, 24
Sheyenne River, Dak. Terr., 53, 205
Shoal, 62, 300
Sierra Nevada, 167
Signal fires, 296
Selenite, 164
Sillery frappé, 91 & n17
Silver, 17
Sioux City, Iowa, 57, 69, 301
Sioux City Daily Journal: arrest of George Custer, 179–80; crowds at Fort Rice, 73; Frederick Grant's departure, 195–97; journey on the *Katie Kountz*, 69–70; September 20, 307–308; steamboats on the Yellowstone River, 145–46; survey preparation, 73–74
Sioux Falls, S.Dak., 77
Sioux Indians, 20
Sitting Bull, *276*; decision to attack, 237–39; defeat of, 313; Gall and, 28; as militant leader, 20; opposition to the 1871 and 1872 surveys, 23–24
Sixth Infantry, 95, 190, 202
Skylark, Mo., 132–33, 164
Slag, volcanic, 168–69
Slaves, 33
Smallpox, 61–63
Smith, J. Gregory: failure to acquire permits, 29; mismanagement of Northern Pacific, 21, 23; W. Milnor Roberts and, 42
Smithsonian Institution, 85
Snake Creek, 223n6
Snake-root, 99
Socrates, 154–55
Soil quality, 100, 125, 198
South Dakota, 56n21, 61n27, 312n24
Sparrows, 125
Spiderwort, 163
Spotted Eagle, 24
Spotted Tail's Agency, 312 & n24
Springfield, S.Dak., 77
Springfield rifles, 165, 255
Stampede, 132, 179, 257
Stanley, Col. David Sloane, *267*; alcoholism of, 16, 33–34, 151, 177; August 4 fighting, 228–31; August 11 fighting, 237–38, 255–57, 263; background, 32–34; Barrows and, 123, 124; battles with Gall, 27–28; Big Muddy Creek and, 154–58; in combat, 224–25; Custer and, 151–52, 175, 177, 179–81; as expedition head, 16; Indian losses and, 311–12; *Key West* and, 194; march tedium and, 147; memoirs of, 95–96; military background of, 83 & n7; negotiation with Spotted Eagle, 24; "Old Standby" as nickname, 255 & n15;

Stanley, Col. David Sloane (*continued*): overseeing column, 117; quarters of, 172; return of, 302–303, 307–309; Rosser's comments on, 136–37, 175, 177; route choices and, 187; and Sitting Bull, 314; stragglers and, 233

Stanley's Stockade, 151, 185–86, 291, 298

Steamboats: *De Smet*, 62–63; *Far West*, 145–46, 145n16; *Ida Stockdale*, 54 & n17, 54n20, 92; *Josephine*, 145n17, *284*; *Katie Kountz*, 69–70, 69n45; *May Lowry*, 304; *Miner*, 63 & n33, 91; new types of, 79; *Peninah*, 146–47, 190 & n4; *Western*, 60 & n26. See also *Key West* (steamboat)

Steam capstan, 146 & n19

Steindachner, Dr. Franz, 65 & n40, 293–94

Stewart, Alexander T., 79 & n4

St. Louis, Mo., 19

Storms, 119–27, 131–32

St. Paul, Minn., 30, 95, 301

St. Paul Daily Pioneer: description of Bismarck, N.Dak., 72–73; Indian attacks, 108; interview with Custer and Rosser, 177–78; survey preparations, 71–72

Stratigraphical geography, 160

Stuart, Gen. J. E. B., 31, 43

Sturgis, Col. Samuel D., 30

Sugar, 102, 161n9, 245

Sully, Gen. Alfred, 165 & n13, 168, 301

Sully's Creek, Dak. Terr., 175

Sunday Creek, 201, 223, 234

Supply depot, 63–64, 173

Surgeons, 62, 237

Surveying, description of, 88–91

Surveyors, 43, 93, 143

Surveyors' reunion, 30n27, 268

Surveys. *See* Yellowstone Surveying expedition (1871); Yellowstone Surveying Expedition (1872); Yellowstone Surveying Expedition (1873)

Sutlers, 99

Swan Lake, 62

Sweet Briar Creek (North Fork Heart River), 107, 164, 311

Tacoma, Wash., 313

Tarpaulins, 156, 299

Taylor, Zach (surveyor), 93

Teamsters, 90, 93, 130, 175

Tedium, 122, 147–48

Teepee, 160–61

Tenderfoot, 44

Tents, dog, 304 & n13

Tents, shelter, 161 & n7

Terry, Brig. Gen. Alfred H., 51 & n13, 79, 301

Theodolites, 90n15

Thomas, Annie, 138n12

Thomas, Gen. George H., 33

Thomas, Theodore, 118 & n3

Thorne, Lt. R. M., 190

Thunderstorms, 119, 120–22, 147

Timber, 125

Tobacco, 102, 131, 222

Tongue River: battle of, 221–33; as camp location, 234 & n26; map of battle at, *218*; as starting point, 48

Tornados, 120–21, 196–97

Townsend, Maj. Edwin: escort of engineers, 105; hailstorms and, 131–32; Indian attacks and, 97 & n2

Tracers, 161

Traders, Indians and, 240

Tragedy, Greek, 313–14

Transcontinental railroads, 16–18, 315

Transit men, 90 & n15, 92

Trappers, 161n9, 189, 311n21

Trash talk, 27–28, 247, 250

Travois (Indian tool), 240 & n4

Trial lines, 89–90

Trinity steeple, New York City, 193

Turnpike, 42n3

Tuttle, Pvt. John H., 226 & n10, 250

Twenty-Second Infantry, 95–96

Unepapa Indians, 233. *See also* Hunkpapa Sioux

Union Army, 31

United States Geological Survey, 297n4

United States Quartermaster's Department, 85, 154

Upland Plover, 125

INDEX

Upper Missouri River, *22*
Utah, 17, 18, 100

Valley City, N.Dak., 53
Varnum, Lt. Charles A., 224 & n8
Vermillion, S.Dak., 77
Vermont Central Railroad, 21
Vermont Clique, 21
Vestal, Stanley, 28 & n24
Veterinary surgeon, 222–23, 234, 259
Victoria Bridge, Wales, 158 & n6
Vienna, Austria, 44, 149, 182n38
Villages, Indian, 235, 237n1, 240–42
Vinatieri, Adam, 254n14
Vinatieri, Felix, 254 & n14
Vinegar, 102, 108, 299
Volcanic slag, 168–69
Volcanoes, 168–70, 181
Volley fire, 27, 247, 254
Vultures, 68, 181

Wages, 204–205
Wagon master, 93
Wagon train, 109–10, 167–68, *289–90*
Walla Walla, Wash., 18
Wall Street, 292
Warp (boating technique), 193
War paint, 241
Washington Territory, 84
Water denudation, 166
Western (steamboat), 60 & n26
Weston, Lieut., 251, 254
West Point, 31, 43, 91
Wetlands, 21
Wheat, 100, 164
Wheelwrights, 110, 149 & n20
Whiskey, 51, 74, 185
Whistler, Maj. Joseph, 136n11, 187
Whistler's Crossing, 136 & n11, 143
White Clay (agency), 312
Whitestone Hill, battle of, 20
Wicke, Iver, 93
Wilderness, Battle of the, 177
Wind, 53–54, 120–21, 126–27
Winding rope, 146 & n19

Windlasses, 193
Windom, Sen. William (Minn.), 93, 203 & n19
Wind River, 33 & n36, 244 & n7
Winona College, 271
Winston, Fendall G., 92 & n21
Winston, Philip B., 51 & n14, 92
Winston, Thomas W., 92 & n21
Wolf Rapids, Mont. Terr., 201
Wolves, 241, 257
Worcestershire sauce, 180
Wounded Knee Massacre, 223n6, 224n8
Wright, Joseph J. B. (surgeon), 33

Yankton, S.Dak., 77–78
Yanktonai Agency, 61
Yarrow, 133
Yates, Capt. George W.: cavalry command of, 165; engineer escort of, 230 & n18; during Indian attacks, 251, 254; marriage to Annie Roberts, 42
Yawl (rowboat), 146
Yellow asters, 124
Yellow Bluffs, 193–94
Yellowstone Park: arrival at, 206; beauty of, 172–73, 189–90; creation of, 18; Jay Cooke and, 19; scientific mysteries of, 30
Yellowstone River, *25, 86, 248–49, 252–53*; boat construction on, 46; crossing of, 244–46; description of, 182, 193–94; flow rate, 193; fording difficulties, 47; length of, 192–93, 192n6; men swimming in, 261; steamers on, 145
Yellowstone Surveying Expedition (1871), 23–24, 162–63
Yellowstone Surveying Expedition (1872), 24–27, 302
Yellowstone Surveying Expedition (1873), 24–27, 302
Yellowstone Valley, 23–24, 233–34
Young Men's Christian Association (YMCA), 207
Yucca, 163–64